# UNNATURAL SELECTION

Also by Mark Roeder:

*The Big Mo: Why Momentum Now Rules Our World*

# UNNATURAL SELECTION

## Why the Geeks Will Inherit the Earth

# Mark Roeder

Arcade Publishing • New York

First North American Edition

Arcade Publishing books may be purchased in bulk at special discounts for
sales promotion, corporate gifts, fund-raising, or educational purposes. Special
editions can also be created to specifications. For details, contact the Special
Sales Department, Arcade Publishing, 307 West 36th Street, 11th Floor, New
York, NY 10018 or arcade@skyhorsepublishing.com.

Arcade Publishing® is a registered trademark of Skyhorse Publishing,
Inc.®, a Delaware corporation.

Visit our website at www.arcadepub.com.
Visit the author's website at www.markroeder.net

10 9 8 7 6 5 4 3 2 1

Library of Congress Cataloging-in-Publication Data
Roeder, Mark.
  Unnatural selection : why the geeks will inherit the earth / Mark Roeder. —
First North American edition.
      pages cm
  Includes bibliographical references.
  ISBN 978-1-62872-435-6 (alk. paper)
  1. Computers—Social aspects. 2. Geeks (Computer enthusiasts)
3. Human evolution I. Title.
  QA76.9.C66R63 2014
  303.48'34—dc23                                    2014020002

Cover design by Anthony Morais
Cover photo credit: Thinkstock

Ebook ISBN: 978-1-62872-480-6

Printed in the United States of America

*For Orlando and his generation*

# Contents

# Introduction:
# The Gift of Weakness

There is a scene at the beginning of the Academy Award–winning film *The Social Network* where a Harvard student declares, "I'm six feet five, 220 [pounds] and there's two of me!" His name is Tyler Winklevoss and he is reassuring his twin brother, Cameron, that he is fully capable of annihilating a fellow student named Mark Zuckerberg, whom he accuses of stealing their idea to launch Facebook. The Winklevosses are a formidable pair—the products of a wealthy family; members of the US Olympic rowing team; and endowed with the chiselled Nordic features of modern-day Vikings. They are the sort of people not to mess with. Mark Zuckerberg, on the other hand, is relatively feeble, highly strung, and appears to suffer from an inability to maintain normal social relations. The odds are stacked against him.

By the end of the film, however, it is Zuckerberg who has soundly defeated the Winklevosses by retaining full control of Facebook, which becomes a runaway global success and transforms the way that over 1.3 billion people communicate and relate to each other. In the process, Zuckerberg becomes extraordinarily wealthy—the youngest billionaire in the world—and a household name.

In many ways, the story of Mark Zuckerberg versus the Winklevoss twins is an ancient one—a classic tale of David and

Goliath. But it's also a brand-new story with a twist. For the real "hero" of *The Social Network* film is not Zuckerberg, but something in the background and barely noticed. This hero is the actual environment in which the Harvard students are now operating—a world that has been transformed by digital communications. Every time the camera peeks into a dormitory or lecture theater or library, we see students glued to computer screens; checking their email, surfing the web or Googling. The university is a far cry from the bookish hallowed halls of yesteryear. It is now an Ivy League Internet café.

It was this new environment that enabled someone like Mark Zuckerberg to succeed against the odds by turning his weakness into strength, and his disadvantages into competitive advantages. He didn't need family connections, or wealth or anyone's permission to succeed; he could act freely without restraint. Conversely, this environment conspired against the more traditional strengths of the Winklevosses. The world had been turned upside down. The geek won.

In evolutionary terms, it would appear that this is an example of how humans—in this case geeks—respond to a new environment. But it's not that simple. For most of human history, our environment was driven by "natural" forces and governed by the laws of the jungle. It was an environment of predators, capricious weather, harsh conditions, scarce resources and a lack of shelter. "These natural forces have shaped our history," explained Professor Iain Stewart, the renowned British geologist. In this primitive unforgiving world, the most valuable human traits related to physical strength, resistance to disease, endurance, animal cunning, resilience and brute force.[1]

More recently, however, the environment has become a lot less natural and is increasingly dominated by man-made influences. This new world is comprised of sprawling, urbanized infrastructure, underpinned by a vast interconnected network of technologies. We spend much of our time immersed within this highly automated, digitized network, glued to screens of all shapes and sizes, often in air-conditioned comfort, at home, in the office, or even while commuting. For many of us, the closest we come to experiencing nature is through watching a wildlife documentary on television. "The nature of Nature has changed," says Rob Brooks, professor of evolutionary biology at the University of New South Wales.[2]

In geological terms we have entered a new epoch called the Anthropocene, or Age of Man, in which, for the first time in human history, the most critical factor shaping our destiny is not the "natural" world but the "man-made" world. Whereas the previous Holocene epoch—which lasted 11,700 years—was relatively stable in terms of man's impact on the Earth and its climate, this new age is the opposite. We now live in a world utterly transformed by the forces of industrialization, urbanization and technology. Dr. Jan Zalasiewicz of the University of Leicester, a leading researcher on the Anthropocene, explained to me, "Our planet no longer functions the way it did. The climate, atmosphere, oceans and ecosystems are all moving beyond Holocene norms. Humanity is driving the Earth across an epoch boundary into some new, geologically novel state."[3]

This book, *Unnatural Selection*, proposes that this new environment is not just affecting our physical world; it is also changing the way we are evolving as a species by immersing us in a "technological greenhouse" that favors characteristics and traits different from previous times. So many people who were once disadvantaged can now bloom; conversely, others who were previously gifted may now struggle; thus enabling, say, a

Zuckerberg to triumph over a Winklevoss. Indeed, this transformation challenges some of our deepest beliefs about what it takes to survive and prosper. It may no longer be simply a case of "survival of the fittest," because in the new epoch, it will sometimes be the least "fit" who will thrive. Their weakness can become a strength.

In fact, we now know that there are many common conditions and disabilities that, in the right environment, can bestow people with superior abilities. Consider, for example, dyslexia, which is the inability to read words accurately and is also known as "word blindness." It afflicts millions of people worldwide, and sufferers usually face a lifetime of learning difficulty. But not always.

Some ingenious experiments conducted by two cognitive scientists at the Massachusetts Institute of Technology, Gadi Geiger and Jerome Lettvin, suggest that people with dyslexia often have perceptual skills that are superior to normal readers. In a *New York Times* article titled "The Upside of Dyslexia," Annie Murphy Paul described how the scientists used a mechanical device for measuring visual comprehension, called a tachistoscope, to show that dyslexics are better at perceiving peripheral fields of vision, and can therefore rapidly take in a scene as a whole.[4] They "absorb the visual gist," which makes them better at interpreting large quantities of data and picking up patterns. This can give them an edge in certain high-tech professions. For example, when a group of astronomers were shown radio signatures—graphs of radio wave emissions from outer space—those with dyslexia could outperform their non-dyslexic colleagues in identifying the distinctive characteristics of black holes.

Paul also cites some intriguing research by Dr. Catya von Karolyi, associate professor of psychology at the University of

Wisconsin Eau Claire, who showed that people with dyslexia can be remarkably good at detecting visual flaws or anomalies. When von Karolyi presented a dyslexic group with a series of Escher-style images—with water flowing upwards or stairs leading nowhere—they were quicker to identify the implausibility of the picture than those without the condition. They saw it more accurately and completely. Writing in the journal *Brain and Language*, Dr. von Karolyi and her coauthors state that "[T]he compelling implication of this finding is that dyslexia should not be characterized only by deficit, but also by talent."

A similar observation can be made about a growing list of conditions that we now know can have a silver lining, including various forms of autism, Asperger's syndrome and attention deficit hyperactivity disorder (ADHD). In certain circumstances, these conditions can actually enhance cognitive performance.

This behavioral transformation corresponds to a provocative new theory called the "orchid and dandelion" theory, or the "sensitivity hypothesis."[5] It suggests that most people are like dandelions, in that they are relatively robust and tend to thrive anywhere. But a small percentage of people are highly sensitive, like orchids. Such people may suffer from an innate affliction, and could even perish without intensive care. Even if they do survive, they are likely to end up as vulnerable, antisocial or dysfunctional members of society.

But put an orchid personality in the right environment (a greenhouse) and they will not just thrive, they will outperform the hardiest dandelion. The orchid and dandelion theory suggests that some of the very genes and tendencies that have long been considered vulnerable and problematic for our species may, in certain environments, be the key to our long-term survival. This does not just apply at the individual level, but also collectively—so entire groups of people (or even cultures) may suddenly find themselves propelled to the forefront because their innate capacities—or so-called weaknesses—begin to resonate with a new environment.

The Anthropocene is such an environment. It represents a kind of "digital greenhouse" that is fostering the development of people whose distinctive traits resonate with the new technological zeitgeist. These technocentric beings are called geeks.

Many years ago I cowrote a book, *How to Beat Space Invaders*, about a popular video game.[6] I was struck by how this early computer game could transmogrify even the feeblest person into an alien-smashing Genghis Khan. But when the geek stepped out into the real world, his newfound self-esteem would evaporate. Nowadays he doesn't really *step outside* anymore because the world has become so technologically integrated that the lines between reality and virtuality have blurred and he can remain Genghis Khan. Such cyberspace conquerors can be found on every financial trading floor today, fearlessly zapping billions of dollars between accounts and making governments tremble at their every move. Like a Space Invaders player, they are insulated from their actions through a virtual interface that maximizes the illusion of control while minimizing the negative consequences. Within this bubble—or techno-greenhouse—these Genghis Khan traders can act with impunity and a heightened sense of self-empowerment.

Not every geek is a Genghis Khan, of course. Most are quiet, cerebral types—"orchids"—who are intensely focused on their work. What they usually have in common, though, is that they thrive in the high-tech environment of the Anthropocene. They are naturally adapted to it and, indeed, are creatures of it.

According to evolutionary psychologists, our behavior—like our physiology—evolves through the process of natural selection. Each new environment can lead to psychological adaptations, which, over time, become heritable traits. So our behavior today is largely driven by the psychological adaptations that our ancestors made a long time ago. These traits are hardwired into us, much like computer operating code, and can be switched on or off in different environments.

It is not a level playing field, however, as the Mark Zuckerberg story shows. Some people are blessed with traits that are highly compatible with the Anthropocene, enabling them to see trends, patterns and opportunities that most of us miss. They may have an uncanny ability to sift through oceans of data and make intuitive sense of them. Or they can overcome conceptual obstacles that hold others back. If they are truly blessed, they may have the "obsessive" trait that compels them to focus intensely on the pursuit of a singular vision until they succeed.

I am talking here, of course, about the geeks.

Geeks, in various guises, have always been part of the human population. But they have tended to operate on the fringes of society and occupy small niches. For example, an ancient Sumerian living in Mesopotamia 2,500 years ago with a knack for numbers may have spent his lifetime etching multiplication tables onto clay tablets. Or a Benedictine monk in the Middle Ages with an exceptional memory may have toiled for years in the monastery cloisters, learning to recite liturgical texts by heart. Moreover, in the past, most people with a prodigious intellectual talent never got a chance to prove themselves, because they had little opportunity for an education, let alone a stimulating career.

It is different today. Acutely bright people are rarely blocked from education, nor are they confined to operating in narrow "niches." In fact, what was considered a niche occupation just a few decades ago—such as software design—has gone mainstream and is now a huge global industry. Yesterday's Sumerian clay etcher can now become a programmer who also owns the company.

Indeed, many of our largest and most influential companies today—such as Google, Facebook, Microsoft, and Apple—are founded and run by geeks. And these famous examples are just the tip of the iceberg. In practically every industry, geeks have emerged from the basements of the corporate hierarchy, where they once dwelled in research departments or laboratories, with chalk on their coats. Now they stride the corridors of power attired in casual gear and sneakers. The geeks now call the shots. What is particularly surprising, however, is how quickly they have evolved to occupy such a dominant position.

Ever since Darwin published his *On the Origin of Species* in 1859, it has been understood that life evolves at a glacially slow pace, over thousands, or even millions, of years. It has been assumed that this gradual process also applies to human behavior and psychology—that it takes a long time to make a psychological adaptation and for it to become an inherited trait. Indeed, the late evolutionary biologist Stephen Jay Gould suggested that humans stopped evolving about 50,000 years ago and that everything we have done since—as a civilization— "we've built with the same body and brain."[7]

But recent discoveries by geneticists have shown that evolution can happen rapidly—at warp speed, in geological terms. For example, a number of species have been found to have evolved significantly over just a few years. These species include stickleback fish in Alaskan lakes, cane toads in Australia, cichlid fish in East Africa, and crickets in Hawaii. Scientists have also discovered that certain human traits can be rapidly inherited from one generation to the next—through epigenetic pathways—a concept that was considered unthinkable just a few years ago.

According to Michael Kinnison of the University of Maine, "Very fast evolution can occur in very short periods." He told *New Scientist* that after he and his colleagues examined the growing list of species that has evolved in recent times, they "started to realize that this (rapid change) was not the exception, it was the norm." The fossil record may not show it, because many of these changes last just a few generations, and therefore won't show up in the geological time scales. But it is becoming increasingly clear that sudden changes in the environment can produce fast-track evolution.[8]

The speed of human evolution is being further accelerated by the explosion in population growth, from an estimated five million people in 9,000 BC to seven billion people today. More people mean more chances that mutations will creep into the human genome. According to anthropologist Gregory Cochran of the University of Utah, member of a team which recently analyzed 3.9 million human DNA sequences, "We found very many human genes undergoing selection," resulting in variation. "Most are very recent, so much so that the rate of human evolution over the past few thousand years is far greater than it has been over the past few million years."[9]

The advent of the Anthropocene, together with recent breakthroughs in our understanding of genetics, shows how quickly we humans are able to respond to environmental changes. Some years ago, Dr. Marcus Pembrey, a prominent geneticist at University College London, wrote a paper about epigenetics in which he asked, "What if the environmental pressures and social changes of the industrial age had become so powerful that evolution had begun to demand that our genes respond faster? What if our DNA now had to react not over many generations and millions of years but within just a few generations?"[10]

This appears to be what is happening today. The Anthropocene is pushing us to evolve faster so that we can cope with the pressures—and opportunities—of a man-made environment. It is favoring traits that enhance our cognitive, emotional and behavioral capabilities—our "neurological software"—for these are the most malleable and responsive to the new environment, unlike our physical characteristics, which take millennia to evolve. Put simply, those people whose neurological software resonates with the Anthropocene will do best, thus enabling, say, a geek to prevail over an old-style industrialist, or an "orchid" to prevail over a dandelion.

This is what this book, *Unnatural Selection*, is about. It is an exploration of how the Anthropocene is encouraging— demanding—us to become "smarter." In evolutionary terms, it encourages us to optimize our brain's neocortex—that area responsible for the higher levels of thinking and reasoning, and which is the most recent part of the human brain to develop. This increased cognitive capacity will enable us to accelerate the technological revolution and urge our species forward. Indeed, some of us are already in the driver's seat of this revolution and have our feet firmly planted on the accelerator. These technocentric beings—geeks—are the personification of the Anthropocene.

The rise of the geek, however, is just the first phase of a broader cognitive revolution. Emerging breakthroughs in the fields of neuropharmacology, nanotechnology and genetic engineering will have a profound impact on the capacity of our minds and how we use them. So-called smart pills, for example, are already being widely used by students, executives, entrepreneurs and academics—to give them an edge in a highly competitive world.

These cognitive enhancement technologies raise important questions about human evolution, because, like medical interventions such as assisted reproduction (IVF), they directly interfere with the process of natural selection. The critical thing about natural selection is that it operates "blindly"; it doesn't push evolution in any particular direction. It simply selects for traits that give an advantage in a particular environment and operates with a mindless, motiveless mechanicity. The evolutionary biologist Richard Dawkins once compared natural selection to a "blind watchmaker" who creates infinite complexity but without a conscious plan.[11]

But increasingly, as humankind interferes with evolution—both directly through medical intervention and indirectly by reshaping the environment—*we* are, in effect, becoming the blind watchmaker. It is early yet, of course. But the sooner we grapple with the possibility of self-determined evolution, the better prepared we will be to deal with it. At the very least, we should encourage the nascent "watchmaker" inside us to take a peek at our work in progress.

The book comprises six main chapters: Chapter 1: "The Anthropocene" explores the advent of the Age of Man and the implications of living in a technological greenhouse. Chapter 2: "Unnatural Selection" looks at how the new environment is selecting for different traits from previous generations, and is accelerating our cognitive evolution. Chapter 3: "The Rise of the Geek" looks at the various types of people—geeks—who are benefitting from the "technological greenhousing" of our world. It considers the impact of such people on society; the challenges they face; and even the threats that they present. For as we shall see, there can be serious consequences when geeks take charge. Chapter 4: "The Cognitive Revolution" looks beyond the rise of the geek and explores emerging technologies that will "enhance" our cognitive and physical capabilities, such as "smart pills," neural implants, and synthetic biology. These tools of unnatural selection will, in effect, enable practically everyone to become a geek, and further accelerate our evolution. In Chapter 5: "Transcendence" I examine how the convergence of various enhancement technologies will result in a more transcendent—or transhumanist—perspective of mankind, whereby we overcome many of the biological limitations on our species. Finally, in Chapter 6: "Backlash," I consider the consequences of bypassing many of the evolutionary safeguards that have protected humans for thousands of years, and explore the growing backlash against some of the new cognitive enhancement technologies and the specter of a transhumanist society.

The book draws on the latest research in technology, genetics, anthropology, biology and cognitive science in an attempt to explain how the Anthropocene environment is reshaping us—individually and collectively. Some of the insights and interpretations expressed in this book, by others, and myself, are controversial, which is not surprising, given the nature of the subject matter. I should also stress that I have not attempted to predict the future—always a misguided task. Rather, I have tried to pull together a range of seemingly disparate themes into a coherent context so that we can better understand the direction in which we are heading.

I hope this book will be of interest to anyone who is curious—or concerned—about the impact of technology on our lives. For we have entered what is likely to be the most challenging period in human history—the Anthropocene—or Age of Man. It means that many of the old rules about survival and success no longer apply. Instead, our destiny will be increasingly determined by the forces of unnatural selection. Those who resonate best with the new technological zeitgeist are very likely to inherit the Earth.

# UNNATURAL
# SELECTION

# Chapter One

# *The Anthropocene*

## *Life, but not as we knew it*

When we think of planet Earth, most of us are conditioned to imagine a luminous blue ball spinning in space, set against an inky darkness. This image has been imprinted on our brains through countless photos, posters and documentaries. Many of us have it as our computer home page, or access it through Google Earth. Indeed, this picture is now so familiar that it has an eternal quality about it—as if it has always been available to us as part of our collective psyche.

Yet it wasn't until 1968 that humans got their first view of Earth in its entirety from space. It was Christmas Eve and the astronauts aboard NASA's Apollo 8 mission had just completed their fourth orbit of the moon when Commander Frank Borman exclaimed, "Oh my God! Look at that picture over here!" The crew peered out from their tiny module and witnessed Earth rising majestically over the lunar horizon. They scrambled for the Hasselblad and snapped one hundred and fifty photos in black and white and color. Of these shots, NASA image number AS8–14–2383 would become one of the most celebrated and ubiquitous images of all time.[1] It is known as Earthrise and presents our planet as an exquisite jewel floating above the stark barrenness of the moon's cratered surface. This photo would change forever our view of home, and as such, our place in the universe.

These days, we take this revelatory image for granted. Newer, more detailed versions of the Earthrise photo can be found practically everywhere. They enable us to view the outlines of the continents, and the vast expanses of oceans, capped by white poles. They show us the contours of mountain ranges, deserts, rainforests and mighty river systems glinting beneath the sun's rays. Such luminous images provide a God's-eye view of our natural world.

What they don't show, however, is the footprint of humankind on our planet. For this we have to turn on the lights.

The first time I saw a night sky view of Earth from space I was shocked by how bright the lights were in the major population centers. I thought it must be an exaggeration—an enhancement to make the map more interesting. Surely, our cities could not possibly emit and consume so much energy. Vast swathes of Europe, Asia and the Americas were bathed in contiguous pools of white light, suggesting our planet was a celestial glitter ball. But this particular image I saw was real and accurate. It was created by the NASA-based Defense Meteorological Satellite Program from an altitude of 500 miles (800 kilometers), and known as the Night Lights map. More recent versions have been compiled from photos taken from the International Space Station, orbiting 240 miles (390 kilometers) above Earth at 17,500 miles (28,000 kilometers) an hour. They are of such high resolution that it is possible to differentiate between various types of street lighting, indicating specific neighborhoods.

These night images provide a graphic picture of humanity's impact on the planet. In fact, if we compare older versions from 2002 to the more recent images—a period of just twelve years—the difference is remarkable and disturbing. Vast tracts of land in developing countries such as China, India and Brazil, that were previously dark, are now lit up. Each of these lights represents a town, city, factory, road, airport or bridge—the relentless sprawl of urbanization and industrialization—and their brightness gives an indication of energy consumption. The stark message from these images is that the most important factor influencing our planet today is humankind, which now consists of over seven billion of us. Together, we have modified over 75 percent of the Earth's land surface to accommodate our needs.[2] Indeed, for the first time in Earth's history, the dominant forces shaping our planet are no longer natural, they are man-made. For this reason, a growing number of geologists have decreed that we have entered a new epoch called the Anthropocene (or Age of Man), and have left behind the relative stability of the Holocene of the last 11,700 years. The advent of the Anthropocene, declared *The Economist* in May 2011, "is one of those moments when a scientific realization, like Copernicus grasping that the Earth goes round the sun, could fundamentally change people's view of things far beyond science."[3]

Despite the significance of this milestone, however, there has been no big announcement or fanfare. We eased into the Anthropocene, as gently and as quietly as slipping into a warm bath. One reason for this absence of ceremony is a lack of clarity about when this new epoch actually began. Some scientists argue that the Anthropocene kicked off with the Industrial Revolution about three hundred years ago, when machinery and large-scale agriculture began to reshape the landscape. Others say the Anthropocene began much earlier—about 8,000 years ago, when humans started farming to supplement their hunter-gatherer subsistence.

There is another view—to which I subscribe—that the Anthropocene commenced as recently as a few decades ago, because this is when the cumulative impact of human activity really started to affect the environment. During this period, known as the Great Acceleration, nearly every indicator of human activity increased exponentially. For example, in just fifty years the global population grew from three billion to over seven billion, most of whom now live in urban areas; economic activity multiplied fifteen-fold; the number of cars increased from forty million to almost one billion; air travel and cargo volumes have risen rapidly; and energy consumption has increased nearly fifty-fold. Initially, the Great Acceleration was driven by Western countries, but as the Night Lights images indicate, it has been boosted by growth in the larger developing countries, including China, India, Brazil and Indonesia.

All this man-made activity has dramatically increased the emission of gases such as methane ($CH_4$) and carbon dioxide ($CO_2$), which topped 400 parts per million in May 2013, the highest level in over three million years.[4] This greenhouse effect traps heat within the Earth's atmosphere and raises temperatures. According to figures from NASA's Goddard Institute for Space Studies—which monitors temperature fluctuations at thousands of sites around the world—the decade beginning January 2000 was the warmest decade since modern records began in 1880.

These rising temperatures are already affecting the Arctic Circle, where the sea ice has shrunk to its smallest extent ever recorded. Satellite images taken in 2012 showed that the area of frozen sea is 50 percent smaller than four decades earlier. Michael Mann, the director of Pennsylvania State University's Earth System Science Center, told the *Guardian* newspaper that that current melting trends indicate sea ice is "declining faster than the models predict . . . The measurements that are coming in tell us it [the melting] is already happening decades ahead of schedule."⁵ Problems like melting ice don't just affect polar bears, it affects all life. "The polar bear is us," said Patricia Romero Lankao, a coauthor of the 2014 UN Climate Panel Report. These increased temperatures are creating a cascading effect that influences not just weather patterns, but everything within the Earth's biosphere. This is the habitable zone in which all life exists, and stretches from the bottom of the oceans to the highest mountains. It is very thin relative to our planet's size—comparable to, say, the skin of an apple—and it bruises easily, as we are discovering.

In effect, what our world is experiencing is a "phase transition," as it is known in thermodynamics, whereby gradual changes converge to create a complete change of state.

## A telluric force

One of the first people to describe the phenomenon of humankind's impact on the Earth was the Italian geologist and Catholic priest Antonio Stoppani, who, in his three-volume *Corso di Geologia* (1873), wrote, "I do not hesitate in proclaiming the Anthropozoic era." In his view, humans represent a "new telluric force, which in power and universality may be compared to the greater forces of the earth. . . . [Man] has become dominant over inorganic matter and over forces that alone had governed him for innumerable centuries."⁶

Stoppani was referring to the Industrial Revolution, which was well and truly under way in Europe at the time. A few decades later, in 1907, the celebrated French philosopher Henri Bergson wrote in his prescient book *L'Evolution Créatrice*, "A century has elapsed since the invention of the steam engine, and we are only just beginning to feel the depths of the shock it gave us. . . . In thousands of years, when, seen from the distance, only the broad lines of the present age will still be visible, our wars and our revolutions will count for little, even supposing they are remembered at all; but the steam engine, and the procession of inventions of every kind that accompanied it, will perhaps be spoken of as we speak of the bronze or of the chipped stone of pre-historic times: it will serve to define an age."

It wasn't until the end of the twentieth century, though, that the actual term "Anthropocene" was coined by a Dutch chemist named Paul Crutzen, who had won a Nobel Prize for helping to discover the hole in the ozone layer. He recalled attending a conference in the late 1990s where "the conference chairman kept referring to the Holocene. I suddenly thought this was wrong. The world has changed too much. So I blurted out: 'No, we are in the Anthropocene' [from the Greek *anthropos*—meaning "human"—and *cene* meaning "new"]. I just made up the word on the spur of the moment. Everyone was shocked. But it seems to have stuck."[7]

Over the past decade, the concept of the Anthropocene has gained a lot of traction. In February 2008, a group of scientists proposed that the term "Anthropocene" be officially acknowledged by the Stratigraphy Commission of the Geological Society of London, which decides when one geological epoch ends and the next one begins. The scientists wrote in their submission, "From the beginning of the Industrial Revolution to the present day, global human population has climbed rapidly from under a billion to its current 6.5 billion [now seven billion], and it continues to rise. The exploitation of coal, oil, and gas in particular has enabled planet-wide industrialization, construction, and mass transport, the ensuing changes encompassing a wide variety of phenomena." These factors have ensured that "we have entered a distinctive phase of Earth's evolution that satisfies geologists' criteria for its recognition as a distinctive stratigraphic unit, to which the name Anthropocene has already been informally given."[8]

As with most things geological, the deliberations of the Stratigraphy Commission will take some time. The process of naming a new epoch is rigorous, and involves numerous working parties, peer-reviewed papers and, given the natural scepticism of geologists, much persuasion. So it may be a few years before the Anthropocene officially joins the ranks of other formally named periods, such as the Jurassic, Pleistocene and Carboniferous. This hasn't stopped many scientists and organizations from using the term. In 2011 the Geological Society of America titled its annual meeting: Archean to Anthropocene: The past is the key to the future.

Yet, despite the growing acceptance of the Anthropocene in the scientific community, the idea has barely made a dent in the public consciousness. This is understandable, given that geological timeframes are not renowned for their sex appeal or newsworthiness. It's not like moving from, say, one musical era to another—from Madonna to Lady Gaga. It is also difficult to explain geological concepts that span thousands, or millions, of years in a twenty-second media sound bite.

Another reason for this lack of awareness is that as a society we are so deeply immersed in—and seduced by—the technological marvels of our man-made world that it is difficult to step outside the system to comprehend their impact. Indeed, this perceptual blindness is one of the defining features of the Anthropocene. We don't clearly see the new world we are now living in because we have helped to create it—it is the water we swim in.

Although we embarked on this journey a long time ago— when our hominid ancestors shaped their first stone tools—it is only very recently that we started to shape the environment itself on a global scale. "By the 21st century," wrote Brett Evans in *Inside Story*, "intelligent life [humans] had become as powerful a force upon the Earth as photosynthesis, or the movement of tectonic plates." This transformation will be accelerated by a rapidly increasing global population, which is forecast to grow from seven billion to 9.2 billion people by 2050. That's another two billion people who will need food, water, energy and resources to live, which, in turn, means more industrialization, urbanization, mining, construction and deforestation. The cumulative impact of all this activity is that our planet is being reengineered on a massive scale.

In the realm of science fiction, there is a name for the process whereby humans transform the ecosystem of another planet. It is "terraforming." The concept was first described in 1942 by author Jack Williamson in a story called "Collision Orbit," and has been subsequently popularized by the astronomer Carl Sagan and others. In our own solar system, there are two planets considered to be candidates for terraforming—Mars and Venus—and possibly some of the moons of the bigger planets, such as Jupiter's Europa. NASA has even conducted research to determine how such planetary transformations could be accomplished. But for the time being, massive projects like this are beyond the scope of current technology and economic capacity, and therefore not viable.[9] Terraforming outer space is something for the future—and science fiction buffs.

But here's the thing. Terraforming is already well under way—on our own planet. We just don't recognize it as such. The difference between terraforming Mars and terraforming Earth is that on Mars we would do it intentionally—according to a plan—but here on Earth we are doing it unintentionally on an ad-hoc basis, almost by accident. And as the Night Lights pictures show, this massive transformation is occurring extraordinarily fast.

## The digital greenhouse

Paradoxically, the biggest terraforming project on Earth is actually the least visible one from space. Even though it began just three decades ago, this project now encompasses the entire globe and affects nearly every inhabitant.

I am talking here about the creation of the global communications network, which acts like a digital nervous system for our planet. The backbone of this vast system is a matrix of high-speed broadband connections, computer servers, fiber-optic cables, and satellite technologies, which feed into millions of wireless towers and antennae located practically everywhere—on the tops of apartment buildings, schools, hospitals and churches. This wireless network connects to over seven billion mobile phones—one for every man, woman and child on the planet—and to nearly a billion screen-based devices such as computers, tablets and other devices.

By the year 2016, it is projected that the amount of information traffic flowing through this network will exceed 900 exabytes—or 900,000,000,000,000,000,000 separate bytes of data—which, if stored on a stack of CD-ROMs, would reach beyond the moon[10]. The rate of growth is staggering. Consider, for example, that the very first email was sent at the Massachusetts Institute of Technology in 1965. In 2006 there were over fifty billion emails sent, and in 2013, over four hundred fifty billion.[11] Our world now generates more data—that is, published and distributed information—in just two days than it created in the entire period of civilization up until 2003.

To cope with this exponentially increasing volume of data, governments and corporations have to constantly upgrade their data storage capabilities. The US Department of Defense, for example, is expanding its Global Communications Grid—which underpins its worldwide capabilities—to handle yottabytes of data. A yottabyte is a septillion bytes, which is the numeral 1 with twenty-four zeros after it. It is a number so large that there is, as yet, no term for a higher order of magnitude. Given the immense scale of the world's communication network, it must rank as one of the greatest engineering achievements of all time. Not surprisingly, this network also generates a lot of radiation—so much so that we now live, work and sleep in an invisible soup of electromagnetic radiation, which is many times stronger than the natural fields in which living cells have evolved over the past 3.8 billion years.

In effect, what we have created is a giant digital greenhouse that overlays the natural physical world. This new simulated landscape—or digital-scape—may not be a land of sweeping plains, bubbling rivers, dense jungles, arid deserts, flora and fauna—but it is real nonetheless, with its own operating principles and consequences. It is the virtual landscape of the Anthropocene.

As we shall see in this book, the new digital landscape may exert an even bigger impact on us than the changes to the physical environment, such as industrialization and climate change. For it has the power to transform the way we experience our world and how we interact with each other. Indeed, it is already doing so. Social media such as Facebook and Twitter are rewiring the neural pathways of community discourse, and enhancing global emotional synchrony by spreading moods from one place to another. Search engines such as Google are revolutionizing how we process information. Online distribution networks such as Amazon and eBay are remodeling consumer behavior.

To cope with this dynamic, data-rich environment, many of us learn to scan vast amounts of information with astonishing dexterity, moving seamlessly between different media such as Twitter, email, telephone, blogs, text messages, television and radio. We have evolved into media omnivores with a voracious appetite for information. Consider that in 2012, the average American consumed three times more information per day than they did in 1980. Similar consumption trends are evident in other developed countries, fuelled by ever increasing sales of iPads and smartphones. Many young people, in particular, are constantly consulting their gadgets for new texts or images—driven by FOMO (Fear of Missing Out) and FONK (Fear of Not Knowing). "We're no longer people of the book," declared Kevin Kelly, the founder of *Wired*. "We've become people of the screen." The Anthropocene, through its digital interface, is redefining our relationship with reality.[12] As if to drive home the point, when thousands of people turned up on Rome's piazza of St. Peter in March 2013 to witness the arrival of the new Pope Francis, there was a mass of screens as far as the eye could see—as people captured and experienced the moment with their smartphones and tablet computers.

Although these developments are exciting—even thrilling— for a society that has become addicted to technology, there is a price to be paid for living life through a screen. Larry Rosen, a psychologist and leading authority on technology overuse, told the *Sunday Age* that future generations will increasingly suffer from "iDisorders"—psychiatric conditions such as narcissistic personality disorder, mania, and attention deficits sparked by excessive use of social media, smartphones and computers. "[Technology] encourages rapid, continuous task-switching, which means that we are only processing information at a shallow level and not deeply so we're not able to have complex thoughts but only superficial ones," he said.[13]

A similar view was articulated a few years ago by Nicholas Carr, who wrote an influential article in the *Atlantic* titled "Is Google Making Us Stupid?" in which he described how the Internet was affecting his own brain. "What the Net seems to be doing is chipping away my capacity for concentration and contemplation. My mind now expects to take in information the way the Net distributes it: in a swiftly moving stream of particles. Once I was a scuba diver in the sea of words. Now I zip along the surface like a guy on a jet ski." Carr cited the playwright Richard Foreman, who suggested that the new digital technologies are encouraging us to become "pancake people," who spread our attention wide and thin, and are incapable of thinking deeply—or dwelling for too long—on any particular issue.[14]

There is no doubt that many people are being forced to dwell in the shallows of the Information Age, and the only way they can cope with so much data is, as Carr suggested, by skimming along the surface. But, as we shall see, not every-one is responding to the pressures of the Anthropocene in this way. Some people are able to simultaneously "skim and dive" the digital oceans and thrive in this information-rich environ-ment. They don't just navigate the net with a jet ski; they also use a submarine. Google is not making these people stupid, but smarter. I am talking about the geeks, whose unique traits ensure they are perfectly at home in the digital world.

Indeed, the very concept of "home" is no longer what it used to be.

## Home is where the hub is

For many of us, our most private and perhaps most sacred space is our home. This might be a house, apartment, loft, cara-van or even a houseboat. It will have walls, a roof and windows, and usually have access to electricity and water. In other words, home is a physical place that occupies certain dimensions in time and space. It has been this way ever since our earliest ancestors huddled together in caves. But this ancient idea of home is changing.

The virtual technologies of the Anthropocene are augment-ing and even supplanting our conceptions of place and space. Increasingly, we spend much of our time—whether working, relaxing or interacting—connected to the Internet, in cyber-space. In 2012, Americans spent an average of twenty-three hours per week online—that's almost one-fifth of their waking hours. The rates are similar in the UK, Western Europe and Australia, as well as in the big cities of China and India. A 2013 study by Cisco Systems on the Internet habits of the world's

Gen-Y population—those under thirty—found that devices such as smartphones "drive every facet of their lives."

Much of this online activity, particularly for young people, involves visiting and updating their "home" pages on social media sites like Facebook. These cyberspace "homes" are not just web addresses or superficial display pages in digital shantytowns. They are serious, well-constructed and meticulously maintained places that people inhabit in the online world. Indeed, for many users, their personal home page has become their home of choice. It provides them with more social interaction, control, freedom, enjoyment, and in some cases, more intimacy than their experience in the physical world. Paradoxically, they feel they can be more authentically themselves in this virtual world. As twenty-five-year-old Dilan Ozdemir, an aspiring teacher, explained to the *Sydney Morning Herald*, "I probably feel more comfortable online, it's just easier to have conversations that you normally wouldn't have in a social setting."[15]

For many people today, "home" is a digital Twilight Zone between the real and the virtual world. But this is no ordinary Twilight Zone, for it is not like being in limbo. On the contrary, it enables its inhabitants to have a clear sense of their bearings and to exert more control over their environment.

Aleks Krotoski, writing in the *Guardian* newspaper, explained it this way:

> Over the last twenty years, we have been encouraged to think of spaces on the web as our homes, from infinitely adaptable personal home pages that we decorate like the walls of a teenager's bedroom, to readymade web hubs such as Facebook in which we surround ourselves with

people and properties that are meaningful to us. In two decades of web research, countless studies have recounted the ways people create environments that signal belonging and identity using text and multimedia in the same way as DIY junkies use paintbrushes and plasterboard. . . . What the web has inspired, then, is a postmodern understanding of what "home" is: a de-physicalized, conceptual and psychological phenomenon . . . the web is another castle that the Englishman can live in, and he seeks to create virtual places that have as much effect on pride, self-esteem and identity as the bricks and mortar version where he sleeps.[16]

Meanwhile, the boundaries between the physical and virtual worlds have become blurred, as many of us operate in both dimensions simultaneously. It is now common for people to use their iPad while watching TV, or to chat on Skype while preparing dinner, or surf the net on their smartphone while flicking through a magazine, or shop online between juggling chores.

Welcome to the anthropocentric version of "home," where bricks and mortar are transmogrified into a digitized semblance of the physical world. Unlike the physical world, however, which appeals primarily to our senses and feelings, this new digital landscape appeals more to our cognitive functions— those intelligent traits that created the Anthropocene in the first place. This means we are interacting with an environment of our own making, in our own image, so to speak. We are no longer just adapters to the environment in a Darwinian sense, we are the shapers of it—and actually responsible for it. "This century is special in the Earth's history," explained Martin Rees of the Royal Society, Britain's Academy of Sciences. "It is the first when one species—ours—has the planet's future in its hands."[17] Or, as the eco-innovator Stewart Brand put it, "We are as gods and have to get good at it."[18]

There are some people, however, who are destined to become particularly good at being gods in the Anthropocene. For as we shall see, they have an *unnatural* advantage.

# Chapter Two

## *Unnatural Selection*

*The difficulty lies not in the new ideas, but in escaping from the old ones.*

—John Maynard Keynes,
Cambridge, 1906

At the beginning of the Industrial Revolution in England, much of the countryside around Manchester was covered in black soot from the new coal-burning factories. Those trees that didn't die from sulfur dioxide emissions were blackened by pollution, making them appear as silhouettes against the bleak sky.

The light-colored pepper moths which had previously camouflaged themselves against the pale bark of the originally light-colored trees now found themselves easy prey for birds. They stood out like luminescent snacks against the blackened bark. The only moths that survived were those that had been fortunate enough to be born a darker shade. These reproduced in such great numbers that by the late nineteenth century, the proportion of dark-colored moths in the Manchester area was 95 percent, a remarkable turnaround from a few decades earlier when the vast majority of moths were white.[1]

The story of the peppered moth of Manchester (*Biston betularia*) is often cited as a classic example of Charles Darwin's concept of natural selection in action. It demonstrates in black and white terms, so to speak, how nature ruthlessly sifts out — or deselects — attributes that are no longer useful in a new environment and supports those that confer a reproductive advantage. It also appears to be the first documented example of how new technology affects evolution.

In the fourth chapter of his 1859 work, *On the Origin of Species*, Darwin explained how the natural selection process works: ". . . if variations useful to any organic being do occur, assuredly individuals thus characterized will have the best chance of being preserved in the struggle for life; and from the strong principle of inheritance they will tend to produce offspring similarly characterized. This principle of preservation, I have called, for the sake of brevity, Natural Selection."

In other words, each new generation of offspring produces new combinations of genes, some of which may enhance the organism's chance of survival and reproductive success. Every so often there is a genetic mutation that can prove useful in certain environments, and therefore may eventually become more common in the population.

Darwin wasn't the first person to suggest that the environment can influence a creature's chance of survival. Nearly 2,500 years ago, the Greek philosopher Empedocles speculated that the origins and development of life were affected by external forces he called the Four Elements (fire, air, earth and water). The ninth-century Islamic writer Al-Jahiz proposed that animals could develop new characteristics to help them in their struggle for existence. Another scholar, Anu Rayhan Biruni, echoed a similar view in the eleventh century. Indeed, throughout recorded history, man has contemplated on the nature of nature, and how it affects us.

But it wasn't until the eighteenth century that interest in the biological sciences really took off, propelled by advances in scientific methods and new technologies. A notable pioneer in this area was the French mathematician and philosopher Pierre Louis Maupertuis (1698–1759), who delved into concepts of heredity and proposed a theory of reproduction whereby life developed according to a "self-organizing" principle.

It was, however, Darwin's own grandfather, Erasmus Darwin (1731–1802), who arguably made the greatest contribution during this formative period. He wrote a book called *Zoonomia*, in which he foreshadowed the modern theory of evolution. In one extraordinarily prescient passage, he proposed, "Would it be too bold to imagine, that . . . all warm-blooded animals have arisen from one living filament, which the great First Cause endued with animality, with the power of acquiring new parts, attended with new propensities, directed by irritations, sensations, volitions, and associations; and thus possessing the faculty of continuing to improve by its own inherent activity, and of delivering down those improvements by generation to its posterity, world without end!" Erasmus Darwin went on to explain how a species propagated itself: "The strongest and most active animal should propagate the species, which should thence become improved." This idea is, of course, remarkably similar to the concept of natural selection.

Notwithstanding the remarkable achievement of his grandfather, it was Charles Darwin who published the first coherent theory of evolution (although Alfred Russel Wallace had concurrently developed an almost identical theory). The validity of this theory has been subsequently confirmed by numerous breakthroughs in biology and genetics. The first of these was Ronald Fisher's work in the 1930s, which showed that the rules of inheritance discovered by Gregor Mendel (through his famous pea experiments in the 1860s) supported Darwinian selection. Almost a century after *Origin*'s publication, in 1953 James Watson and Francis Crick announced their discovery of DNA, the genetic building block of life. This prompted further advances in molecular biology which helped pioneering evolutionary biologists Ronald A. Fisher, Sewall Wright, J. B. S. Haldane, Julian Huxley, Theodosius Dobzhansky and others to develop what is known today as the modern synthesis of evolution—the cornerstone of which is natural selection.

It is difficult to overstate the importance of Darwin's ideas, not just to science, but also to the wider world. According to evolutionary biologist Stephen Jay Gould, the radicalism of natural selection lay in its power to "dethrone some of the deepest and most traditional comforts of Western thought." It challenged the long-standing belief that humankind held an exalted place in the natural world, as deemed by a benevolent creator. The philosopher Daniel Dennett called it "Darwin's dangerous idea," and the concept of evolution by natural selection "a universal acid which cannot be kept restricted to any vessel or container, as it soon leaks out, working its way into ever-wider surroundings."[2]

Today, Darwin's ideas influence—and permeate—many disciplines that underpin our modern civilization, including economics, politics, culture, humanities, international relations, ideology and psychology. The term "survival of the fittest" has become synonymous with Darwinism (even though it was coined by the philosopher Herbert Spencer), and is frequently used to rationalize laissez-faire ideologies such as capitalism and deregulation.

Although Darwin never intended his ideas to be used in a way that justified a "law of the jungle" mindset, or to be misconstrued as "social Darwinism," they have taken on a life of their own. The German political philosopher and co-originator of the communist ideology Friedrich Engels wrote in 1872 that "Darwin did not know what a bitter satire he wrote on mankind when he showed that free competition, the struggle for existence, which the economists celebrate as the highest historical achievement, is the normal state of the animal kingdom."

Over the past century, Darwin's ideas have been used by some of the most ruthless regimes the world has known, including the Nazis, who used them to justify their program of eugenics and the creation of a master race of blond-haired, blue-eyed Aryans. In 1940, a future winner of the Nobel Prize in Physiology or Medicine, Konrad Lorenz, used Darwinism to articulate a case for the Nazi state, which he would later disown. He wrote ". . . selection for toughness, heroism, and social utility . . . must be accomplished by some human institution, if mankind, in default of selective factors, is not to be ruined by domestication-induced degeneracy. The racial idea as the basis of our state has already accomplished much in this respect."

More recently, Darwin's ideas have been misconstrued by supporters of creationism and intelligent design in order to promote the idea that all life was created by God or an intelligent being, rather than through evolution. According to Theodosius Dobzhansky, "Their [Creationists'] favourite sport is stringing together quotations, carefully and sometimes expertly taken out of context, to show that nothing is really established or agreed upon among evolutionists. Some of my colleagues and myself have been amused and amazed to read ourselves quoted in a way showing that we are really anti-evolutionists under the skin."

Even Darwin's own words have been taken out of context to suggest he didn't really believe in evolution. There is a quote in his *Origin of Species* which reads:

> To suppose that the eye with all its inimitable contrivances for adjusting the focus to different distances, for admitting different amounts of light, and for the correction of spherical and chromatic aberration, could have been formed by natural selection, seems, *I freely confess, absurd in the highest degree* [emphasis mine].

But the quote doesn't end there; it goes on to read:

> ... Yet reason tells me, that if numerous gradations from a perfect and complex eye to one very imperfect and simple, each grade being useful to its possessor, can be shown to exist; if further, the eye does vary ever so slightly, and the variations be inherited, which is certainly the case; and if any variation or modification in the organ be ever useful to an animal under changing conditions of life, then the difficulty of believing that a perfect and complex eye could be formed by natural selection, though insuperable by our imagination, can hardly be considered real.

In other words, Darwin was in no doubt about the reality of evolution.

Notwithstanding the frequent attacks on Darwinism, his concept of evolution and natural selection remains as valid today as when it was first proposed over one hundred fifty years ago. It provides the only rational and scientific explanation of how life has evolved on our planet over the past 3.8 billion years. As Dobzhansky wrote, "nothing in biology makes sense except in the light of evolution."[3]

# From beaks to brains

When Darwin conceived his theory, he was primarily concerned with the physical evolution of animal species—the size of a bird's beak, the shape of a lizard's claw, or the colors of a moth wing. Although he assumed the rules of natural selection also applied to humans, he was cautious about how they might affect human psychology. Nevertheless, he did predict in *On the Origin of Species* that psychology would develop an evolutionary underpinning: "In the distant future I see open fields for far more important researches. Psychology will be based on a new foundation, that of the necessary acquirement of each mental power and capacity by gradation," he wrote.

Darwin devoted two of his later books to the study of psychology and animal emotions—*The Descent of Man, and Selection in Relation to Sex* (1871), and *The Expression of the Emotions in Man and Animals* (1872). These works helped inspire a century of further research into the biological basis of behavior, which culminated in a series of groundbreaking books, such as the 1975 publication of *Sociobiology* by the Harvard entomologist E. O. Wilson, and the 1992 publication of *The Adapted Mind: Evolutionary Psychology and the Generation of Culture* by Jerome Barkow, Leda Cosmides and John Tooby, which laid out the theoretical foundations of evolutionary psychology.

Today, evolutionary psychology, like natural section, exerts a powerful influence over many dimensions of our world and has applications in fields such as psychology, psychiatry, economics, law, the arts and politics. Its influence has grown stronger in recent years, as breakthroughs in genetics have provided more support for a biological underpinning to the evolution of human behaviors. For example, our capacity for language—the most basic of all cognitive functions—has recently been linked to the FOXP2 gene, which is thought to help us use grammar and articulate sounds (even Neanderthals carried it). Indeed, according to advocates of evolutionary psychology, such as the linguist and philosopher Noam Chomsky and the cognitive scientist Steven Pinker, there is now strong evidence that the human psyche has adapted to changes in the ancestral environment. That is, it is a product of natural selection or sexual selection. They cite examples of behaviors that occur universally in all cultures, such as the ability to read people's emotions, cooperate with others, fear dangerous creatures such as snakes and spiders, identify healthy mates, appreciate beauty, discern intelligence, and distinguish family members from non-family members.[4]

Notwithstanding the powerful case for evolutionary psychology, many scientists and philosophers remain skeptical. They are concerned about reducing human behavior to a mechanical genes-driven perspective that implies a lack of "free will." Some argue that man's creative imagination and spiritual aspirations cannot be explained by natural selection, which is, in effect, an "algorithm" that operates in a mechanical, automatic and impersonal way. Moreover, they believe, the mind is so infinitesimally complex, and there are so many variables involved in the human genetic puzzle, that human psychology can respond to changes in the environment in unpredictable, irrational, and perhaps even non-Darwinian ways.

They point to the fact that although the brain weighs just 1.4 kilograms (3 pounds), it comprises about 85 billion neurons, each of which is linked to about 10,000 other neurons through long, thin extensions called dendrites and axons. In total, the human brain has between 100 trillion and 1,000 trillion neural connections.[5] To further complicate matters, the brain is comprised of thousands of different types of brain cells with names like Betz, Renshaw, Purkinje and Golgi, each of which behaves differently. In contrast, organs like the kidney or lung are much simpler in terms of structure. According to Jeff Lichman, a neuroscientist at Harvard University who is attempting to create the first map of the human brain at a cellular level, it will take the equivalent of one million petabytes of information to complete the project, "which is more than the digital content of the world right now," he says. Given that one petabyte is the equivalent of twenty million four-drawer filing cabinets filled with text, don't expect to see this map any time soon.[6]

The astonishing complexity of the human brain, however, does not preclude the possibility that our psychology can evolve through natural selection. The simplest principles often govern the most complicated structures. But we should at least be open to the possibility that the evolution of the mind may not be as straightforward as, say, changing the colors of the Manchester moths. Indeed, recent research into child behavior suggests that our psychological evolution can sometimes appear to contradict the rules of natural selection, and what constitutes evolutionary "fitness."

## Dandelions and orchids

In 2005, two human development specialists, Bruce J. Ellis of the University of Arizona and W. Thomas Boyce of the University of British Columbia, published a research paper called "Biological Sensitivity to Context" in the journal *Development*

*and Psychopathology.*[7] Their paper looked at children's susceptibility to their family environment, and found that most kids
are relatively robust and tend to survive in nearly all types of
situations. They named these resilient ones "dandelion children" after the hardy flower that grows just about anywhere.
By contrast, the researchers observed, there was another,
smaller group of children who were much more sensitive to
their family environment—particularly the quality of parenting
they received. They branded these kids "orchid children," bor-
rowing from the Swedish idiom *orkidebarn* (*orkide* meaning
"orchid" and *barn* meaning "child").

Predictably, when the sensitive orchid children were neglected
or mistreated, they soon withered because they couldn't cope.
"They sustained higher rates of disease, disorder, and injuries than
their more normatively reactive peers from the same environments," Ellis and Boyce wrote. As they grew up, they were also
far more likely to develop depression, abuse drugs, or go to jail.

But surprisingly, the researchers also discovered that when
orchid childred were adequately nurtured, they not only
survived—they thrived—and often outperformed the dandelion
children. Their heightened sensitivity to their environment enabled them to see things differently, and to respond more crea-
tively to opportunities. "The highly reactive biological profiles
found in this subset of children reveal a unique sensitivity or
'permeability' to the influence of environmental conditions,"
observed Ellis and Boyce. "In conditions of neglect, the orchid
promptly declines, while in conditions of support and nurture, it
is a flower of unusual delicacy and beauty."

This discovery, known as the "sensitivity hypothesis"
or the "dandelion and orchid" theory has profound implications for the way society deals with issues like childhood

delinquency, truancy, substance abuse and learning difficulties. It also has the potential to revolutionize our understanding of how we respond to changes in our environment. Because here we have examples of "undesirable" psychological traits that are not eliminated by nature's ruthless selection process, but are actually preserved. These orchids are just waiting for the right environment to make use of their unique characteristics—for their inner frog to be kissed so it can become a prince.

I should stress, however, that we are talking here about a relatively small subset of orchid children who have the potential for such extraordinary capabilities. There will be many children whose afflictions are so severe that no amount of nurturing will transform their weakness to strength.

So what is it about some orchid children that endows them with such remarkable potential?

Danielle M. Dick, a geneticist at Virginia Commonwealth University, has been studying some orchid children. Together with thirteen other scientists, she has been exploring a gene called CHRM2, which is implicated in alcohol dependency and appears to influence other behavioral disorders. Dick's team took DNA samples from more than four hundred boys and girls from kindergarten to age seventeen, in three US cities, and analyzed variations in their CHRM2 gene. They collected detailed information about the children's behavior—or rather, misbehavior—including aggression, delinquency, drug abuse, and so on. They learned where the kids went in their spare time, who they hung out with, how they spent their money, and whether their parents knew what they were up to. Notwithstanding the real-world difficulties of monitoring children this way, the results did give a reasonable indication of the quality of care provided by the parents, by revealing whether their child was neglected or treated with indifference.[8]

Their findings were published in the April 2011 edition of *Psychological Science*, and suggested that children with a certain variation of the CHRM2 gene will behave in a way consistent with the orchid child model of susceptibility.[9] That is, if they receive negligent parenting, they will become troublesome teenagers and exhibit negative behavior. But if they receive positive and attentive parenting, they will exhibit the most desirable teenager behavior. In other words, the children who were most likely to behave badly in a negligent home environment were also the most likely to do well in a nurturing, healthy home environment.

Given these findings, it would be tempting to call CHRM2 the "orchid gene," but this would be an oversimplification. Behavior is rarely attributable to a single gene, but rather to a combination of genes and environmental influences working synergistically. It is reasonable to assume, though, that the genes of the orchid children help make them highly susceptible to a change in the environment.

The bigger question is, perhaps, why are these genes there in the first place? Why weren't they weeded out by natural selection, given that they make children so hypersensitive and vulnerable? Surely it would be better for the orchid traits to be replaced by the more robust dandelion ones? Isn't that what "survival of the fittest" is all about?

Not really. Natural selection doesn't care which genes stay or go. It operates blindly and expresses no preference for what is useful or not. If a gene or mutation turns out to be advantageous in a particular environment, it may be preserved. But this doesn't mean that other "non-useful" genes—such as the CHRM2 (orchid) gene—will be "deselected" or eliminated. They may turn out to be useful in a different environment. In fact, each of us carries within us an infinitesimally huge variety of dormant genetic traits—most of which will never be activated. This is not evolutionary baggage, but rather an ingenious hedging strategy to ensure that we have sufficient potentiality to deal with a new environment.

Indeed, many of our genes are being constantly switched on or off in response to changes in our environment. In fact, it is not even necessary for the actual environment to change— what matters is that we *perceive* that it does. This is enough to affect the way our genes are expressed, which in turn, affects our behavior. The discovery of this phenomenon has led to the development of a new field called social genomics, which explores how society influences our genes, and vice versa. According to psychological researchers George Slavich and Steven Cole of the University of California, Los Angeles, "As we learn more about human gene transcription [how our genes are expressed], it is becoming increasingly apparent that our molecular 'selves' are far more fluid and permeable to social-environmental influence than we have generally appreciated."[10]

This flexibility means that we, as a species, can adapt and learn more easily. So, for example, if we find ourselves in an intellectually challenging environment, our dormant cognitive traits will respond accordingly, as certain genes are switched on or off. And some people—due to their innate potential—will respond more effectively than others.

This process was first described over a century ago, in 1896, by the pioneering American psychologist James Mark Baldwin (1861–1934). In his paper "A New Factor in Evolution," he suggested that some people are born with a gift for learning certain things, such as linguistic dexterity, which gives them an advantage. Over time, this ability may be aspired to, and emulated by, more people, and become a "sustained behavior." This makes it easier for those with the learning trait to reproduce because they are perceived as, dare I say it, "sexy" (in the same way that geeks are today). Eventually, the learning trait is assimilated into the wider population and becomes part of its genetic repertoire.

According to the renowned philosopher of science Daniel Dennett, Baldwinism enables the "speeding up of the basic, slow process of natural selection" by operating as an evolutionary "crane" which gives a species a lift up or head start.[11] So, just as sexual reproduction gave humans more possibilities than asexual reproduction did, by mixing up our chromosomes and providing more variety—so too does Baldwinian evolution, by enabling learning traits to rapidly develop in a population. This means we have more intellectual flexibility to be able to survive and reproduce under conditions that are as wide and varied as possible. It opens up more options.

Which brings us to the issue of "neurodiversity" and its critical relevance to the Anthropocene.

## Neurodiversity

During the early 1990s, when social media began to take off, a group of people in the United States with a high functioning form of autism began to meet online and express their frustration at being stigmatized because of their neurological condition.[12] They felt they were being treated as second-class citizens. They believed the core of the problem was that the medical and psychiatric establishment held a very narrow of view of what constituted a "normal" person, and that anyone who was not "neurotypical" was abnormal and had to be treated accordingly. These early activists saw this negative stereotyping as a human rights issue and began lobbying for the acceptance of a more tolerant and inclusive perspective of neurological disorders. In essence, they argued that society should accept neurological diversity in the same way it accepts biodiversity in the natural world.

In 1998, an article appeared in the *Atlantic* by Harvey Blume in which he wrote:

"In trying to come to terms with an NT [neurotypical] dominated world, autistics are neither willing nor able to give up their own customs. Instead, they are proposing a new social compact, one emphasizing neurological pluralism. . . . The consensus emerging from the Internet forums and Web sites where autistics congregate . . . is that NT is only one of many neurological configurations—the dominant one certainly, but not necessarily the best."[13]

In effect, Blume was making a case for "neurodiversity," a term that was originally coined by Australian autism advocate Judith Singer, who is on the Asperger's spectrum.

Over the past decade the neurodiversity movement has expanded beyond autism to include a range of conditions including dyslexia, dyspraxia, Tourette syndrome, speech disorders, bipolar disorder, ADHD, schizophrenia, and circadian rhythm disorders. And the list keeps growing.

The bigger point is, though, that it is not just people with these conditions who suffer as a result of society's narrow view of what a "normal" person is. We all do.

In the 1970s, there was a popular TV series called *The Brady Bunch*, which featured an American family comprised of beautiful children and loving parents. I recall wishing that my own family, with all its idiosyncrasies, could be more like the idealized version I saw on television. That is, a "normal" family. The Brady parents—Mike and Carol—were the personification of a loving married couple who, together with their loyal housekeeper, Alice, presided over a family where every conflict was resolved amicably—just in time for the final commercial break. For many viewers, the star of the show was the fresh-faced Marcia Brady, who demonstrated what a mature, well-balanced teenager was supposed to be.

Years later, I learned that the real life Marcia was anything but. The actor who played her—Maureen McCormick— revealed in her autobiography, *Here's the Story: Surviving Marcia Brady and Finding My True Voice*, that behind the smiles was a darker story. She was battling cocaine addiction, trading sex for drugs, had two abortions and was partying at the Playboy mansion. "I was hiding the reality of my life behind the unreal perfection of Marcia Brady," she wrote. "No one suspected the fear that gnawed at me even as I lent my voice to the chorus of Bradys singing, 'It's a Sunshine Day.'"

McCormick's personal problems were exacerbated by the unrealistic expectations she had to live up to as part of America's perfect family, in which everyone seemed to be so well-adjusted and normal.

And this is the problem.

Modern society, despite its liberal aspirations, has a narrow definition of what "normal" means in a behavioral sense. Everyone is expected to comply with a rather strict range of parameters and exhibit "neurotypical" behavior. Anyone who steps too far outside this narrow frame is considered abnormal, and therefore in need of psychological treatment. In recent years, the list of behaviors considered to be abnormal has grown exponentially, as psychologists discover—or invent—more types of disorders. Many of these are legitimate afflictions, while others are more dubious. This diagnosis creep has been exacerbated by pharmaceutical companies, which are always searching for new conditions—or markets—to develop drugs for. Increasingly we see conditions that were once considered normal labeled as disorders, such as sadness as depression; or joy as hysteria; or enthusiasm as mania; or energetic behavior as hyperactivity, and so on—all of which can be treated with a drug.

Christopher Lane, the author of *Shyness: How Normal Behavior Became a Sickness*, explains, "Before you sell a drug, you have to sell the disease." He says that the de facto marketing manual for psychiatric illnesses is the *Diagnostic and Statistical Manual of Mental Disorders* (DSM for short), which is produced by the American Psychiatric Association. Over the past few decades this manual has expanded from a thin spiral-bound handbook into a hefty tome, as more disorders have been added. The most recent version of the handbook, DSM-5—which was released in 2013—caused a backlash around the world among health professionals, largely because of "diagnosis creep."

The British Psychological Society, representing over 50,000 members, wrote of the DSM-5 preliminary draft: "We are concerned that clients and the general public are negatively affected by the continued and continuous medicalization of their natural and normal responses to their experiences. . . . We believe that classifying these problems as 'illnesses' misses the relational context of problems. . . . The putative diagnoses presented in DSM-5 are clearly based largely on social *norms* [my emphasis], . . . the [diagnostic] criteria are not value-free, but rather reflect current *normative* [emphasis mine] social expectations."[14]

It also warned that the proposed inclusion of attenuated psychosis syndrome "looks like an opportunity to stigmatize eccentric people, and to lower the threshold for achieving a diagnosis of psychosis [thus] increasing the number of people receiving antipsychotic medication." While some may view the Society's response as a veiled defense of British eccentricity, it does highlight the dangers of diagnosis creep.

This discrimination against certain behavioral types is not limited to parts of the medical establishment. There is a deep and pervasive bias in our society against some basic psychological characteristics. For example, in our media-dominated era it is no longer okay to be an "introvert." Everyone is expected to be an extrovert, or possibly even a finalist on *Britain's Got Talent*. According to Susan Cain, writing in the *Guardian* on March 14, 2012: "We live with a value system that I call the Extrovert Ideal—the omnipresent belief that the ideal self is gregarious, alpha and comfortable in the spotlight. The archetypal extrovert prefers action to contemplation, risk-taking to heed-taking, certainty to doubt. He or she favours quick decisions, even at the risk of being wrong; works well in teams and socialises in groups."

Cain says that although we like to think that we value individuality, all too often we admire one type of individual—the kind who is comfortable "putting themself out there." Research suggests that the vast majority of teachers believe that the ideal student is an extrovert. As adults, many of us work for organizations that insist we work in teams, in offices without walls, for supervisors who value "people skills" above all. To advance our careers, Cain says, "[W]e're expected to promote ourselves unabashedly. The scientists whose research gets funded often have confident, perhaps overconfident, personalities. The artists whose work adorns the walls of contemporary museums strike impressive poses at gallery openings. The authors whose books get published—once a reclusive breed—are now vetted by publicists to make sure they're talk show ready."

Yet, there is no correlation between extroversion and worldly success. "Some of our greatest ideas, art, and inventions—from the theory of evolution to Van Gogh's *Sunflowers* to the personal computer—came from quiet and cerebral people who knew how to tune into their inner worlds and the treasures to be found there," Cain says.

Without introverts, the world would be devoid of Newton's theory of gravity, Einstein's theory of relativity, W. B. Yeats's "The Second Coming," Chopin's Nocturnes, Proust's *In Search of Lost Time*, Peter Pan, Orwell's *Nineteen Eighty-Four*, *The Cat in the Hat*, Charlie Brown, the films of Steven Spielberg, Google (cofounded by introvert Larry Page) and Harry Potter.

Despite this anomaly, Cain notes that:

"Introversion—along with its cousins sensitivity, seriousness, and shyness—is now a second-class personality trait, somewhere between a disappointment and a pathology. . . . Extroversion is an enormously appealing personality style, but we've turned it into an oppressive standard to which most of us feel we must conform."

The gradual narrowing of psychological "norms" can be traced back to the birth of modern psychology when, in 1590, a German philosopher, Rudolph Goclenius, created the word "psychology"—from the Greek words psyche, "soul," and logos, "study." It was around this time that the first hospitals for mental patients were established in Europe, initially intended for only the most extreme cases of mental disorders. The province of the psyche was still largely in the hands of the Church, which was more interested in preserving moral codes than in ensuring psychological consistency. As long you were a true believer of the Faith, and weren't a "witch," it was okay to be a bit weird. It took another three centuries before psychology was accepted as a legitimate independent field when, in 1879, Wilhelm Wundt established the first laboratory dedicated exclusively to psychological research at Leipzig University in Germany. Wundt subsequently became known as the "father of psychology," and helped inspire many others including the American philosopher William James, the memory researcher Hermann Ebbinghaus, and the Russian physiologist Ivan Pavlov, who is renowned for his conditioning experiments in dogs.

The most famous of all the early cartographers of the human psyche is Sigmund Freud, who developed a particular method of psychotherapy called psychoanalysis. He used clinical and interpretive approaches—including dream analysis— for tackling taboo subjects such as sexuality, repression and the unconscious mind. He also coined the terms "id," "ego" and "superego," and inspired other luminaries such as the Swiss psychiatrist Carl Jung. Although these pioneers continue to exert a strong influence over theory and practices, the field of psychology has developed in many new directions over the past century, including behaviorism, humanism, Gestalt, existentialism and cognitivism, to name just a few. Despite this great diversity of psychological approaches, however, they are all aiming for the same goal. That is, to return the patient to a "normal" state of functioning, as far as is practicable.

Indeed, this is the great paradox of modern psychology; by pushing society toward the idea of a healthy "norm," it is undermining the neurological diversity that society needs to stay healthy—not to mention the individual rights of those afflicted. For, as we have seen with the orchid children and people with high functioning forms of conditions such as Asperger's, dyslexia and ADHD, it is sometimes the so-called non-normal people (non-neurotypical) who can be the biggest contributors to society. Neurodiversity is vital to our collective survival. It is nature's hedging strategy to ensure that our species has a broad range of psychological traits—both favorable and non-favorable—to ensure it can deal with any environment. These negative traits are not always weeded out by natural selection; they are often preserved. As Harvey Blume wrote in the *Atlantic*, "Neurodiversity may be every bit as crucial for the human race as biodiversity is for life in general. Who can say what form of wiring will prove best at any given moment? Cybernetics and computer culture, for example, may favor a somewhat autistic cast of mind."[15]

Indeed, we now know that some people with autism can often outperform so-called normal people at tasks that involve processing complex visual information and large data sets—the very skills that underpin the Internet age. According to Laurent Mottron, a professor of psychiatry who holds the Marcel and Rolande Gosselin Research Chair in Cognitive Neuroscience of Autism at the University of Montreal, Canada:

> In certain settings autism can be an advantage. A growing body of research shows that autistics outperform neurologically typical children and adults in a wide range of perception tasks . . . and most autistic people outperform other individuals in auditory tasks (such as discriminating sound pitches), detecting visual structures and mentally manipulating complex three dimensional shapes. They also do better in Raven's Matrices, a classic intelligence test in which subjects use analytical skills to complete an ongoing visual pattern.[16]

There is also a high concentration of autism among techies and elite math students, according to studies conducted by Professor Simon Baron-Cohen (the cousin of actor Sacha Baron-Cohen) at Cambridge University's Autism Research Centre.[17] These innate skills have enabled autistics to become very successful in the technology fields—so much so that there would probably be no Silicon Valley without them.

One of the first people to make this connection was *Wired*'s Steve Silberman, who, in an influential article titled "The Geek Syndrome," wrote, "It's a familiar joke in the industry that many of the hard-core programmers in IT strongholds like Intel, Adobe, and Silicon Graphics—coming to work early, leaving late, sucking down Big Gulps in their cubicles while they code for hours—are residing somewhere in Asperger's domain." Within this unique environment "their autistic minds soar in the virtual realms of mathematics, symbols and code." They can also act with more freedom because "[i]n the geek warrens of engineering and R&D, social graces are beside the point. You can be as off-the-wall as you want to be, but if your code is bulletproof, no one's going to point out that you've been wearing the same shirt for two weeks."[18]

Not all people on the autism spectrum lack social graces, of course, but many do—and they are often described as having a lack of empathy, as highlighted by Baron-Cohen's research.[19] This tag is disputed, however, by autism researchers such as Damian Milton, who suggest that empathy is a "two-way street" that requires both parties to appreciate each other's perspective.[20]

Nevertheless, given their social differences, individuals with autism will not thrive in all careers and will often struggle in people-oriented fields, such as retail or customer service. Despite these caveats, Laurent Mottron believes that autistics can make a huge contribution to society. "The hardest part is finding them the right job," he says, one that creates an environment in which they can truly bloom and exploit their innate cognitive gifts.

Which brings us back to the Anthropocene.

The world we have now entered, with all its whizz-bang technology, has created a digital greenhouse that is ideal for nurturing neurodiversity. Perhaps for the first time in human history, and despite the efforts of the psychological establishment, this new environment can actually favor the non-neurotypical person, whether they are hypersensitive, fragile, introverted, antisocial, or afflicted with autism, ADHD, Asperger's, or any number of conditions considered to be abnormal. This is not to suggest that these conditions—in certain situations—may not have been helpful in our evolutionary past. But rather that the balance has shifted significantly more in their favor. In effect, the Anthropocene emulates Clark Kent's phone box, whereby the mild-mannered "introverted" reporter walks in and steps out again as Superman. Consider for a moment how different are the high-tech tycoons of today—such as the founders of Facebook, Google, Wikipedia, Microsoft and Twitter—from the typical industrialists of the past century. They may share the ingenuity and drive of their twentieth-century forebears, but these super-geeks are a different breed in a behavioral sense. They are more cerebral, introspective and low key—and, indeed, quirky. It is unlikely they would have succeeded to the same extent in the more physical world of the industrial age. No board of directors would have appointed them as CEO to drive the company forward. No banks or investors would have put up the money to get things rolling. It is far more likely that these bright geeks would have found themselves heading up a research department, or an experimental skunk works, safely locked away in their labs.

There were exceptions to this, of course, such as the inventor Thomas Edison—the super-geek of his time—who various biographers have suggested suffered from ADHD but was also a successful businessman.[21] But it is instructive to note that Edison's collaborator, the brilliant electrical engineer Nikola Tesla, who developed the modern alternating electric current, and who was arguably the most famous inventor in America, died penniless and in debt in January 1943. For the most part, geeks did not fare well in the old industrial environment. But now they have stepped out into the bright sunlight, their eyes squinting like Bill Gates behind thick-rimmed glasses.

It is not just neurological diversity, however, that is being encouraged by this new Anthropocentric environment—physical diversity is too. Even a person with a disability may be able to turn it to their advantage.

## Love at first byte

Rollo Carpenter was born deaf in one ear. It wasn't a debilitating condition and he learned to live with it. By the time he started attending his local school in Cranbrook, England, he had become so well adapted to his deaf ear that even his classmates and most of his teachers were unaware of it. "It may well have affected my schoolwork a bit," he recalled, "and it was always difficult to hear in noisy social situations. Nevertheless I was a reasonable student but was never a star at anything."[22]

In 1981, when he was sixteen years old, Rollo's attention was caught by an advertisement in the *Sunday Times*. It was for a small computer called the ZX81 Sinclair, which could be ordered by mail for ninety-nine pounds, the sum of all his savings. The Sinclair was extraordinarily simple by today's standards—it had a mere one kilobyte of memory, the only drive was a standard cassette tape, and the screen was a black-and-white display from an old TV. Yet, so great was its appeal that this little machine gave birth to the mass-market computer industry in Britain. "The minute I saw it I knew I had to have it," said Rollo.

It was love at first byte.

Rollo immersed himself in the micro world of the Sinclair and spent countless hours in his room tinkering, programming and teaching himself how to master his computer. "I even learned to program machine code," he said, referring to a relatively rudimentary form of programming.

Despite the fact that Rollo's attention was now being consumed by the Sinclair, leaving less time for homework—and causing him to stay up far too late—his school performance actually improved. For the first time in his life he found himself getting As. It was as if a light had been switched on inside Rollo's head and he could now operate at a different level, despite his deaf ear. Three years later he won a place at prestigious Oxford University. Over the next two decades, he became one of Britain's leading designers of cutting edge computer programs and artificial intelligence (AI). He invented Cleverbot, a popular application that is able to converse with people in a remarkably realistic fashion, and which has conducted over sixty-five million conversations since it was launched. He won the Loebner Prize in 2005 and 2006 for designing the most human-like computer simulation. In 2010 he won the British Computer Society's Machine Intelligence Competition for his work in AI, and in 2011 his software came close to passing the famous Turing Test, which adjudicates on how "human-like" a computer is.

Rollo explained to me that the key to his success in creating human-like language programs—known as chatterbots—is that his simulations don't just focus on *what* is being said in the present moment, but on the *context* within which the communication is happening. So, for example, his chatterbots pay attention to a wide range of signals and previous responses to develop a "sense" of what is being discussed rather than relying on the literal meaning of each sentence. So even if a piece of information is missing, or misunderstood, there is enough background information to form a coherent response.

This is precisely the approach that Rollo used to overcome his hearing disability; he looked at previous verbal and other nonverbal cues in a situation to complete the communications puzzle and create a meaningful context. When he entered into the world of computers, this contextual ability became an advantage, by enabling him to take a more holistic perspective that transcended literal meanings and linear trains of thought. In other words, what began as a hearing weakness became a strength in the new hi-tech environment of artificial intelligence. If Rollo had been born centuries earlier in, say, Africa, his hearing difficulty would have put him at a distinct disadvantage; he may not have heard the sound of a cracking twig beneath the paw of an approaching lion. But in the Anthropocene, he could turn his weakness into strength in a way that resonated with the technological zeitgeist.

I should emphasize that I am not suggesting that every person with a physical or neurological condition will thrive in the Anthropocene. The majority of people with autism, dyslexia, anxiety, ADHD and severe physical disabilities, will, unfortunately, continue to suffer cruelly from their condition, as will their families and friends. But there will be a proportion of them who are able to capitalize on the new environment and turn their so-called weakness into advantage. Through the process of unnatural selection, these orchids—especially the tech-savvy ones—may not only survive, but thrive.

What this shows is that we, as a species, are far more malleable and adaptable than is commonly understood. Our limitations are not necessarily fixed and immutable, they can be modified, harnessed or even turned into strengths in the right situation. We are like evolutionary alchemists—capable of converting leaden traits into golden ones. But for this alchemy to occur, there must be a constructive interaction between our nature and the environment.

## Nature is nurture

For many years, biologists and evolutionary psychologists have grappled with the question of whether genetics (nature) or the environment (nurture) has more influence over human behavior. The debate has swung back and forth in response to the latest scientific breakthroughs and ideological fashion. The nurture camp has argued that the environment is the critical factor, and can overcome even the most stubborn behavioral traits. Whereas the nature camp says that behavior is hardwired in the genes, and will usually prevail in the end. Traits are "bred in the bone" they say.

We have already seen how critically important the environment is for eliciting certain behaviors—whether they be empowering an orchid personality to bloom or inspiring Mark Zuckerberg to create Facebook. In these cases, the environment appears to act like a fairy's magic wand, transforming the people it touches. But we also know that in all these situations there has been an underlying condition or trait waiting to be activated and brought to life.

What's really happening is a mutual interaction between the environment and a nature, in which neither side is more influential. It takes two to tango. In fact, the nature versus nurture debate is now over. "Both sides won, because they're both vitally important," explained David Shenk, the author of *The Genius in All of Us*. He said that the new science shows us that they do not operate independently. "Declaring that a person gets X percent of his/her intelligence from genes and Y percent from the environment is like saying that X percent of Shakespeare's greatness can be found in his verbs, and Y percent in his adjectives. There is no nature vs. nurture, or nature plus nurture; instead, it's nature interacting with nurture, which is often expressed by scientists as 'GxE' (genes interacting with environment)."

This concept is called "interactionism" and is now supported by an increasing number of geneticists, neuroscientists and psychologists.[23] It has challenged many conventional ideas about human intelligence. For nearly a century, the gold-standard tool for measuring intelligence—particularly in the West—has been the IQ (intelligence quotient) test. When Lewis Terman at Stanford University, California, invented the IQ test in 1916—following earlier work by the European psychologists Alfred Binet and L. Wilhelm Stern—he thought he had invented a way to determine a person's native intelligence. By distilling the complex spectrum of cognitive processes of the human mind into a single number—where 100 represents the average—Terman provided a handy tape measure of intellectual ability for schools, governments, the military or anyone who needed it.

We now know, however, that IQ tests don't tell the whole story—and can be misleading. Firstly, they imply that intelligence is fixed, despite the fact that the average IQ in developed countries has risen by nearly 3 percent per decade since the early 1940s, and recent studies have shown that a person's intelligence can change throughout their lifetime. So, contrary to popular belief, IQ is not a fixed attribute like the color of a person's eyes or their height. It is quite malleable.

We have also witnessed a significant closing of the gaps between different cultural, social and racial groups. For example, in the United States, the IQ gap of eleven points between blacks and whites has been closing, primarily because of environmental changes. It has also been shown that if a child in the United States from a low socio-economic background is given sufficient nurturing at an early age, their IQ will increase significantly. A University of North Carolina study known as the Abecedarian Early Intervention Project provided such disadvantaged children with intensive educational, nutritional and health intervention from infancy to age five. In follow-up tests, these children showed an advantage of six IQ points over a control group, and this improvement persisted decades later. A thirty-year update on the Abecedarian Project published in the project's website in January 2012 showed that those children who received the intervention were four times more likely to have completed a university degree than those who hadn't received the intervention, and they had been more consistently employed. "What's more," said Dr. Elizabeth Pungello, a coauthor of the latest study, "this achievement applied to both boys and girls, an important finding given the current low rate of college graduation for minority males in our country."

These positive changes cannot be explained by genetics because the DNA would not have changed. A more likely explanation is that they are caused by environmental changes which alter the cognitive, and perhaps even the physical, structures of the human brain through a process known as neuroplasticity. Our minds are constantly remodeling themselves in response to the stimuli they receive by creating new neural pathways. "In other words," said Shenk, "intelligence isn't fixed. Intelligence isn't general. Intelligence is not a thing. Instead, intelligence is a dynamic, diffuse, and ongoing process."

Indeed, the very concept of the standardized IQ test is under a cloud because it is biased in favor of people who grew up in the Western culture that created it. This is why some sociologists consider the standard IQ test to be a "measure of our adaptation to modernity" rather than a measure of an innate capability.

Notwithstanding the difficulties in measuring intelligence, there are, of course actual differences in intelligence between people, and we now know that some of these differences can be inherited. Researchers at the University of Manchester studied two types of intelligence in more than 3,500 people from Aberdeen, Edinburgh, Manchester and Newcastle—and examined more than half a million genetic markers in every person. They found that between 40 and 50 percent of the differences in people's intelligence could be traced to genetic factors.[24] But what matters more is how we capitalize on our available intelligence. For example, there is a widely held conception that the reason East Asian students (particularly Chinese) do so well at

American universities is because they are born "smarter." The reality is that they work harder. They embody the Confucian ideals of scholarship through diligent effort and are often seen on campuses congregating in after-hours study groups while most other students have gone home or are playing sports.

The second misleading aspect of IQ tests is that, given the importance we assign to them, they can imply that our prospects in life will be automatically determined by our IQ result. Lewis Terman believed so fervently in this hypothesis that he spent his lifetime trying to prove it.[25] His most famous experiment was a longitudinal study of hundreds of gifted children—known as "Terman's Termites"—whose IQs ranged from 130 to 150 and over. Terman tracked down 730 of them when they had reached middle adulthood to see how well they had fared. He divided them up into three groups—As, Bs and Cs. The As comprised the top 20 percent and had achieved great success in life, having become academics, doctors, engineers, lawyers and so on. The Bs comprised 60 percent of the group and had done reasonably well, with many having completed tertiary education and pursuing successful careers. The remaining Cs, however, had not fared well at all. One-third of them were college dropouts, only eight of them had received university degrees, and many were in low-paid jobs such as postal work or were unemployed. This was a disturbing and puzzling result considering how acutely bright these people had been as children. According to their IQ they should have had the world at their feet.

Terman tried to account for what had happened to the lackluster Cs and explored various explanations, including their mental health, personality characteristics, hobbies, ages when they began walking and talking, and so on. But in the end, it became clear that there was only one significant difference. The Cs came from disadvantaged family backgrounds compared to those of the As and Bs. Unlike their more fortunate peers, the Cs did not grow up surrounded by books and stimulating conversations around the dining table in a nurturing environment. Like the poor children in the control group of the Abecedarian project, they had never experienced a constructive environment. As they progressed into adulthood, despite their precocious intellects, the Cs carried their early conditioning with them and tended to replicate it wherever they went.

Remember, this was before the Internet when the opportunities for cognitive development were more tangible, and therefore inflexible. Universities were physical places of bricks and mortar, with gates at the front. There were no virtual online courses easily accessible through a cheap computer; or search engines, or chat rooms or any of the facilities of the digital biosphere that we have today. Without the advantage of a nurturing environment, the Cs floundered like delicate orchids cast outside the greenhouse. Not even their genetically gifted minds could save them.

The tale of the Termites reminds us how susceptible and responsive we are to the environment around us. This doesn't just apply to individuals, but also to communities, and indeed, to entire nations.

Consider the case of Israel, whose population of 7.2 million is comprised mostly of Ashkenazi Jews. During the Middle Ages, their ancestors living in Europe were, for various cultural reasons, a genetically isolated population that could make a living only in certain intellectually demanding jobs, such as money lending and wealth management. Over a few centuries, it is likely that these pressures—together with the Jewish culture's strong focus on education—accelerated their intellectual development. According to the anthropologists Gregory Cochran and Henry Harpending, authors of *The 10,000 Year Explosion*, the average IQ of the Ashkenazi Jews today is 112, about three-quarters of a standard deviation above the European mean, and they account for a disproportionate number of geniuses relative to the size of their population.[26]

This intellectual advantage has helped the Ashkenazi Jews to do astonishingly well during the second half of the twentieth century, following the tragedy of the Second World War years. As David Brooks wrote in the *New York Times*, even though Jews comprise just 0.2 percent of the world population, they account for 27 percent of the Nobel physics laureates and 31 percent of the medicine laureates. In the United States, where Jews make up 2 percent of the population, they comprise 37 percent of Academy Award-winning directors, 51 percent of Pulitzer Prize winners for non-fiction, 21 percent of Ivy League student bodies, 26 percent of Kennedy Center honorees, and 38 percent of leading philanthropists.[27] The Jewish success story is replicated in other fields too, such as finance, fashion, science, design, technology and the arts.

But as the twentieth century drew to a close, there was one country where the Jews were not doing well. Israel.

During the 1980s and 1990s, the country was languishing economically, politically and culturally. Economic growth had stalled, there was a banking stock crisis, and by 1985 annual inflation had soared to nearly 450 percent and was expected to reach 1000 percent. There was little local investment in scientific research and development. Tensions with neighboring countries were on a knife-edge and immigration to Israel had stalled. These indicators caused foreign investors to stay away, which further exacerbated the negative conditions, which persisted up until the turn of the century. This scenario prompted the economist Milton Friedman to joke that "people used to think Jews were good cooks, good economic managers and bad soldiers; Israel proved them wrong." It was as if the nation's intellectually gifted Termites were all Cs.

All this changed around the turn of the century when Israel underwent a transformation of tectonic proportions, driven largely by the high-technology industry. As Brooks explained:

"Tel Aviv has become one of the world's foremost entrepreneurial hot spots. Israel has more high-tech start-ups per capita than any other nation on Earth, by far. It leads the world in civilian research and development spending per capita. It ranks second behind the U.S. in the number of companies listed on the Nasdaq. Israel, with seven million people, attracts as much venture capital as France and Germany combined."

When Apple Computers decided to launch its first research and development facility outside America, it chose to do so in Haifa and Herzliya Pituah, in Israel, where Apple's CEO, Tim Cook, explained there is some "fantastic technical talent."

What caused this change? After all, we are talking about the same people living in the same country with the same neighbors and beset by the same seemingly intractable geopolitical problems.

The answer is that the environment changed, and radically. A series of bold economic reforms which made it easy to start new businesses and "the arrival of a million Russian immigrants [many of whom are highly educated and ambitious] and the stagnation of the peace process have produced a historic shift," wrote Brooks. "The most resourceful Israelis are going into technology and commerce, not politics. This has had a desultory effect on the nation's public life, but an invigorating one on its economy."

## Geek heaven

In many ways, Israel today is a microcosm of the Anthropocene. The nation has evolved into a high-tech digital greenhouse—a geek heaven. Sure, it has the advantage of a highly educated and innately intelligent workforce. But these advantages have always been there more or less. What really changed was that the new environment began to nurture dormant innovative talents. It switched on the orchids.

This same phenomenon is happening throughout the world today. Twenty years ago there was only one large-scale high-tech hothouse—Silicon Valley in the United States. Now there are dozens. They can be found in places like Shanghai, Beijing, Shenzen and Hong Kong in China; Bangalore, Chennai, Hyderabad and Pune in India; Cambridge and London in the UK; Zurich and Geneva in Switzerland; Tokyo, Japan; Seoul, Korea; Helsinki, Finland; Moscow, Russia; Singapore; Copenhagen, Denmark; and Sydney, Australia.

The interesting thing about nearly all these new hothouses is how they came about. They developed spontaneously. They are not the result of a deliberate effort to create "technology parks" and "high-tech centers of excellence." Such architect-designed places often feel so artificial and sterile that they end up as vacant white elephants. According to Paul Graham, a successful American programmer and technology investor, "A government that asks 'How can we build a Silicon Valley?' has probably ensured failure by the way they framed the question.

You don't build a Silicon Valley; you let one grow."[28] The reason for this is that the type of people who are attracted to high-tech centers are looking for a particular environment that combines a sense of freedom, authenticity, social interaction and character. If you want to attract geeks, said Graham: "[Y]ou need more than a town with personality. You need a town with the right personality."

Graham said that Silicon Valley provided the ideal environment for high-tech start-ups because it was located in the San Francisco Bay Area, which had been a magnet for the young and optimistic for decades before it was associated with technology.

"It was a place people went in search of something new," he said, and so it became synonymous with California nuttiness. There's still a lot of that there. If you wanted to start a new fad—a new way to focus one's "energy," for example, or a new category of things not to eat—the Bay Area would be the place to do it. But a place that tolerates oddness in the search for the new is exactly what you want in a start-up hub, because economically that's what start-ups are. Most good high-tech start-up ideas seem a little crazy; if they were obviously good ideas, someone would have done them already.

Another high-tech hotspot is located in lower Manhattan New York, where dozens of start-ups have created "Silicon Alley," which now challenges Boston as the epicenter of innovation on the East Coast. Budding entrepreneurs come to New York to tap into the concentrated pool of talent, the city's vibrant creative energy, and direct access to the venture capital markets. By April 2014, the New York City technology ecosystem generated nearly $125 billion in annual output, according to a report conducted by HR&A Advisors.[29] Once again, the rise of New York as a high-tech center was not the result of a deliberate government strategy—although Mayor Bloomberg certainly encouraged it—but rather the result of spontaneous growth factors.

It's not just the location that makes a place a digital hothouse, it's also the living arrangements. In recent years there has been a resurgence in communal living among ambitious young tech entrepreneurs. This is not just a matter of saving money before they strike it rich, it's about cross-fertilization of ideas and the opportunity to inspire each other.

Rainbow Mansion is a 460-square-meter commune located in the heart of Silicon Valley, just south of San Francisco. Since it was founded in 2006, this sparsely decorated, salmon-colored building has been home to aspiring tech entrepreneurs from more than twelve countries. They come here to bounce ideas off each other, enjoy the camaraderie and inspire each other in ways that, hopefully, will help them change the world. In an interview with April Dembosky of the *Financial Times*, a thirty-two-year-old Google software engineer named Loredana Afanasiev explained, "When I started living at Rainbow, I started learning more at home than I did at work."[30]

Rainbow was originally founded out of necessity when, in 2006, some new NASA employees needed a place to live, and saw the residence advertised on Craigslist. But it soon began attracting geeks from all around the world keen to join in their regular discussion nights. For example, when Sir Roger Penrose, the English mathematical physicist renowned for his work on general relativity and cosmology, visited Silicon Valley, he stopped off at NASA, Google and the Rainbow Mansion. But he spent most of his time at Rainbow, where he gave a talk in the Rainbow library about black holes, Hawking radiation, and the unitary principle—and spent the night in a Rainbow guest room.

A couple of times a year, Rainbow hosts a "hackathon," during which dozens of computer geeks bring their laptops over and share some beers and pizzas. As Eric Stackpole, an engineering associate at NASA who set up a workshop in the garage to work on his robotic submarine, told Dembosky, "People aren't just living together; we're building ideas that will last for years."

The unconventional hours that people keep also have the effect of blurring work and play, and help spur them on. Damian Madray, a young entrepreneur living at a communal hothouse called Glint (another sprawling 600-square-meter mansion in Twin Peaks, San Francisco), explained to Dembosky: "When you're an entrepreneur, you often work until two a.m. and you feel like you're the only person doing it. At the Glint, everyone is up until one or two a.m. There's a certain comfort and motivation in that—my friends are hacking away at this, I might as well keep hacking."

Beyond the need for cross-fertilization and like-minded spirits, the other key ingredient for technical innovation is a sense of freedom. Paul Graham believes that it is no coincidence that the most technologically creative cities in the United States are also the most liberal. "But it's not because liberals are smarter. It's because liberal cities tolerate odd ideas, and smart people by definition have odd ideas. Conversely, a town that gets praised for being 'solid' or representing 'traditional values' may be a fine place to live, but it's never going to succeed as a start-up hub."

Recall, from the Introduction, how Mark Zuckerberg was able to create Facebook in the liberal environment of Harvard

University. Despite its solid Ivy League credentials, the intellectual atmosphere at Harvard was open and free. But what really made it a hothouse for innovation was the advent of the Internet, which transformed the university into a giant Internet café with a very different set of rules from the "physical" world.

In his review of the Facebook film *The Social Network*, Harvard Professor Lawrence Lessig wrote that the key thing about the Zuckerberg Facebook story, "… is not that he's a boy genius. He plainly is, but many are. It's not that he's a socially clumsy (relative to the Harvard elite) boy genius. Every one of them is. And it's not that he invented an amazing product through hard work and insight that millions love. The history of American entrepreneurism is just that history, told with different technologies at different times and places. Instead, what's important here is that Zuckerberg's genius could be embraced by half-a-billion people within six years of its first being launched, without (and here is the critical bit) asking permission of anyone."[31]

The disruptive technology of the Internet had broken down the old barriers of entry for new ideas by enabling "permission-less" innovation. It created a whole new environment in which people could express their ideas and bring them to life. Their minds were set free.

What's interesting—and unprecedented—is how quickly these ideas have permeated our culture. Concepts like "permission-less innovation," "immediacy," "transparency" and "virtuality" were considered to be aspirational and unrealistic just a few years ago, yet we now take them for granted in the Internet age. It is as if these ideas have been able to mutate, replicate and spread like genes.

Indeed, according to the evolutionary biologist Richard Dawkins, this is exactly what has happened.

# *Mighty memes*

In his book *The Selfish Gene*, Dawkins proposed that ideas and behaviors spread from person to person through "memes," which, he suggested, are the cultural equivalents of genes. Memes could include catchphrases, melodies, fashion and even the technology for building arches. And just like a gene, a meme can self-replicate, mutate and respond to environmental pressures. It is a unit of cultural inheritance. "Memes propagate themselves in the meme pool by leaping from brain to brain via a process which, in the broad sense, can be called imitation," Dawkins wrote. They also compete with each other for our attention—our brain time. Those memes that propagate less prolifically may become extinct, while others may thrive and spread to affect the behavior of a greater number of hosts. Some memes—such as ideologies or stereotypes—can survive even though they are detrimental to the welfare of their hosts, or the broader community.

Memes are also highly adept at capitalizing on any changes in the environment, particularly man-made ones. So, for example, when Gutenberg printed the Bible in 1454—on the world's first printing press—the Christian religious meme "went viral" through the mass dissemination of books.

In today's world of social media, memes have the capacity to spread at near instantaneous speed. They thrive like digital viruses in the vast interconnected network of the Anthropocene, able to penetrate every cell and host. A fertile meme can ignite the imagination of millions of people and topple governments almost overnight, as we have seen in the Middle East. For once an idea, such as freedom, becomes embedded in a meme, it rapidly gains traction and can displace obsolete ideas through a form of "memetic natural selection."

The concept of neurodiversity—discussed earlier—is also a meme, which fosters the tolerance of a greater range of behavioral types—especially that of geeks. By reshaping the cognitive environment, memes affect how our minds evolve. Indeed, the idea of natural selection is itself a meme, Dawkins has suggested.

Memes also shape our perspective of history. Writing in *The Smithsonian*, James Gleick described how, during Isaac Newton's lifetime, barely a few thousand people actually knew what Newton looked like. "Yet now millions of people have quite a clear idea—based on replicas of copies of rather poorly painted portraits. Even more pervasive and indelible are the smile of Mona Lisa, *The Scream* of Edvard Munch and the silhouettes of various fictional extraterrestrials. These are memes, living a life of their own, independent of any physical reality. 'This may not be what George Washington looked like then,' a tour guide was overheard saying of the Gilbert Stuart portrait at the Metropolitan Museum of Art, 'but this is what he looks like now.' Exactly."[32]

It is clear that memes exert a powerful influence on social evolution. Let's now extend this concept further and consider an intriguing possibility. What if certain memes are more likely to resonate with particular cultures, at different times? That is, the memes and cultures work synergistically. This may help explain why certain memes can gain so much traction in some cultures and not in others, because they resonate with cultural traits in the local populations.

To get a sense of how this would work, consider the example of software designers from India. This distinct group has become remarkably successful in recent years. Step into a high-tech environment anywhere in the world and you will usually see Indian computer scientists busy at work developing complex algorithms and codes. They can excel at it. When Bill Gates got serious about offshoring some of Microsoft's capabilities in the 1990s and hiring more talented people, where did he go? India. Likewise for dozens of other major technology firms.

But why?

One reason is that many Indians today have access to a good education system that emphasizes math, and Indian students tend to be highly motivated and diligent. This enables India to churn out thousands of new engineers and computer scientists every year. But so too do China, Germany, South Korea, Japan and many other countries. And yet, they are not over-represented in the global software industry like people from India are. Let's consider another perspective.

Among the educated elite of India there is often—but not always—a tendency to look down on manual labor. It is a legacy of the nation's caste system, which stipulated that only lower classes would perform such work. Today it manifests in subtle ways. For example, an Indian engineer will often emphasize the cognitive nature of his or her work, rather than the practical fieldwork. "A civil engineer is not a mason cutting stones," an Indian engineer told a Western academic, "for he is engaged in intellectual, not manual work."[33] This attitude contrasts sharply with that of Western (and Chinese) engineers, who are often keen to get out into the field, wander around in muddy boots, and earn their stripes as site supervisors. Indeed, even medicine was once considered to be an unclean profession in India (particularly by the Brahmans) because of its association with pollution from corpses and bodily fluids.

The point is that many bright young people in India are not just looking for intellectually stimulating jobs—like their Western counterparts—they are also looking for "clean" jobs.

Computer software design is, of course, an immaculately clean business. It derives solely from the pristine faculties of the mind and deals in artificial constructs. Robin Jeffrey, a visiting professor at the Institute of South Asian Studies in Singapore, and coauthor of *The Great Indian Phone Book*, suggested to me that "computer work has the advantage for a great many people in India because it could be regarded as 'ritually antiseptic' in that you don't have to deal with things that are ritually polluting."

The other important thing about software design is that it requires a unique combination of numeracy, creativity, and linguistic dexterity. And here, once again, many educated Indians have a distinct advantage. This is because they have had to grapple with the complex linguistic structure that underpins their society. India is a hugely diverse nation, and there are literally hundreds of languages, ranging from Indo-Aryan ones such as Hindi, to the various Dravidian languages like Tamil; and, more recently, English, which was introduced under British rule. Within this cacophony of cultural and linguistic influences, a word or nuance can convey a multiplicity of meanings—about caste, status, wealth, power and interrelationships.

This means that language is not just a means of communication for many Indians—it helps them navigate and define their complex social and political landscape. It is little wonder that Indians have developed an acute sensitivity to linguistic structure and meaning—particularly the educated elite.

It is little wonder also that when computer software began to take over the world in the late 1980s, these young Indians were uniquely placed to resonate with the new techno-centric environment. They responded positively to the "software meme," because they intuitively understood that, like Indian languages, software is not just about communication, it delineates structure and process. And it was "clean" work, which appealed to their cultural sensitivities.

Before long, thousands of ambitious, educated young Indians had moved into software programming, and become so successful that they were eagerly sought after by companies like Microsoft, Google, and Facebook.

But this would not have happened before the Anthropocene. It took an entirely new environment to "switch-on" the dormant cultural potential of many of these bright young people. A similar process was described earlier, in terms of social genomics, whereby certain groups of genes are switched on and off in response to changing social conditions. It also has parallels with a theory proposed by the psychologist Donald Hebb, who suggested that when groups of brain cells are consistently stimulated in a certain way, they form bonds with neighboring cells and modify their biological structure so as to promote associative learning. In other words, "neurons that fire together, wire together," thus potentially making them more capable of learning, and more receptive to the kinds of memes that we see at work with the Indian software designers.

Processes such as social genomics, meme transmission and neural bonding may, in the future, ensure that we see individuals in other cultural groups resonate with an emerging technology and become successful. For example, it is likely that computers will evolve beyond using digital transistors to incorporate quantum phenomena such as superposition and entanglement, or nano-biotechnology, which uses DNA and proteins (DNA computing). This will require a mindset more accustomed to dealing with paradoxes and holistic perspectives rather than the linear, rational modes of thinking, which have dominated

Western thought for the past five hundred years—and particularly since the Industrial Revolution. Hence, those cultures that extol a more holistic perspective of life, may find themselves better equipped to deal with a world driven by quantum- or biology-based technology, with its inherent paradoxes and complexities. It may resonate with their ancestral cultural memes and underlying genetics in ways that enable them to interact with the new technology more effectively. Indeed, it is possible that cultures that are today considered to be technologically backward will eventually be the powerhouses of tomorrow—because they resonate better with the emerging technological zeitgeist. These dormant "orchid" cultures are just waiting for the right environment to switch them on.

This brings us to one of the most controversial aspects of evolutionary theory, which was touched on earlier but which deserves further exploration: epigenetics.

## The epigenetics revolution

Until quite recently, it was generally assumed that it took a long time for a species to change in an evolutionary sense. Darwin thought it took tens of thousands, or perhaps millions of years. The fossil records certainly suggest this. After all, natural selection requires many generations to work its magic on the genetic building block of life, DNA.[34]

Recent scientific discoveries, however, show that evolution can occur very quickly—at warp speed, in geological terms. Recall from the Introduction how a number of species have been found around the world to have evolved significantly over just a few years—the cichlid fish in East Africa, crickets in Hawaii, cane toads in Australia, and stickleback fish in Alaska. Phillip Gingerich, a researcher at the University of Michigan, has written, "I think a superficial reading of the fossil record has given us a misleading picture of the evolutionary process. The changes seen over long intervals of geological time are not representative of what happens on a generation-to-generation timescale." He says that a species can actually evolve very quickly.[35]

It's not just the speed of evolution that has taken the scientific community by surprise, but the mechanisms by which it can occur. It has long been thought that for a species to evolve, its DNA must be changed through the process of chance mutations in order for new characteristics to be inherited by the next generation. This DNA-based mechanism is known as "hard inheritance," because the nucleotide sequence of DNA is usually constant, and it can only be changed by rare random mutations that are passed down through the generations. In other words, it is "hard" to do.

But many scientists now believe that there are alternative paths for evolutionary changes to occur, and that they operate much faster than traditional DNA transmission. The most intriguing of these involves epigenetics—from the Greek term *epi* meaning "above or beyond"—which operates like a master control switch altering the way genes are expressed in response to the environment, and thus affecting behavior. Although the exact mechanism is still poorly understood, it is believed that epigenetics involves chemical modifications to the DNA itself, or the proteins with which the DNA associates.

The most surprising revelation, however, is that in certain situations these epigenetic changes may be passed from one generation to the next—a concept that would have been unthinkable a few years ago. For it means that, theoretically at least, the experience an individual has in their own lifetime can influence their offspring through the epigenetic process. This process is known as "soft inheritance" because it doesn't change the DNA—only the way the genes are expressed.

Many biologists remain deeply skeptical about the capacity of epigenetics to affect evolution. They believe that only hard genetic material—DNA—can be passed from generation to generation. What really rankles them is that epigenetics strays dangerously close to an idea considered to be long dead and buried—Lamarckism.

Jean-Baptiste Pierre Antoine de Monet, Chevalier de Lamarck (1744–1829), often known plainly as Lamarck, was a French pre-Darwinian evolutionist who believed that a species can directly modify itself to suit its environment. He wrote, for example, that giraffes have long necks because their ancestors were "obliged to browse on the leaves of trees and to make constant efforts to reach them," and that shore birds acquired long legs by constantly stretching their legs to lift themselves out of the water. Lamarck was not able to explain how this "soft inheritance" worked, and so his theories were subsequently discredited by Darwin, and also by the work of Gregor Mendel, whose experiments on pea plants showed how inheritance actually works.

But recent revelations about the power of epigenetics have prompted scientists to take a second look at Lamarck's theory of soft inheritance. Eric Richards, a professor of biology at Washington University in St. Louis, has analyzed current and past research in epigenetics and the history of evolution, and wrote that epigenetics should be considered a form of soft inheritance, citing examples in both the plant and animal kingdoms.

"From a molecular biology point of view there is a mechanism to do soft inheritance."[36]

That mechanism, Richards has suggested, could operate through a process known as methylation, whereby methyl groups (one carbon and three hydrogen atoms) switch certain genes off or regulate them down, thus affecting the way they are expressed. It may also affect the way DNA is "packaged" in terms of how closely it is wrapped around the proteins, much like threads around a spool. Loosely wrapped DNA is more easily accessed, and therefore more easily expressed than tightly wrapped DNA.

A simple way of thinking about this is that the epigenetic process reduces the "boot-up time" for a change to occur. So rather than having to go through the long route of changing the underlying DNA through random mutation, the change can happen through the way the gene is expressed or embellished. It's an evolutionary shortcut. Richards cites the example of pregnant mice whose diets (environment) were varied in an attempt to change the DNA methylation of their progeny. "The idea was," he explained, "[I]f you pump these pregnant moms up with these dietary supplements, you might be able to skew the DNA methylation patterns, and thus skew the way the mice come out at the end of the day, and it works. In this particular instance, what you're getting fed in the womb influences your phenotype—physical and physiological attributes."

Another study he quoted showed that—similar to the dandelion and orchid studies in humans—when young rats were placed in a nurturing environment, the methylation of their glucocorticoid receptor gene—which affects metabolism and immune responses—gets turned on and is expressed early at a critical period. This makes the pup more able to handle stress later in life. If there is insufficient nurturing and grooming then the gene never gets turned on.

"These studies show that the early nutritional environment in the mice and early behavioral environment in the rat studies can change the DNA packaging on the genome, and that this is 'remembered' in the cell divisions that make the rest of the organism," said Richards.

Richards's work is corroborated by a recent experiment conducted by Dr. Cath Suter, head of the epigenetics laboratory at the Victor Chang Institute in Sydney, and her fellow researchers.[37] They fed some genetically identical mice a special diet that switches off a certain gene. These mice became brown and lean, while the other mice with the gene still active and fed on a different diet became obese and yellow. The researchers found that when the diet was continued in lean mice over five generations, these epigenetic effects were inherited. The proportion of lean, brown mice in each subsequent generation increased, despite no alteration of their genetic code. Suter explained to me that when the diet was withdrawn, there was a reversal effect, but some of the changes persisted—which suggested "there was a progressive, cumulative effect with each new generation." The research also showed that with each new generation of epigenetically affected mice, there was more variability among them even though their underlying DNA remained the same.

Suter has speculated that epigenetics could play a role in rapid changes in the human population, including the obesity epidemic. A mother's diet during pregnancy, she suggested, could influence her child's, and even her grandchild's, propensity to develop metabolic syndrome, and thus increase the risk of cardiovascular disease and diabetes, irrespective of what the child or grandchild eats. "It would seem that we have an ancestral responsibility to our descendants," Suter said, which directly challenges the widely held notion that we can do as we please in our own lifetime, without fear of passing on a legacy to our children and grandchildren. Our behavior has consequences on the biological continuum.

Over the longer term, Suter and her colleagues suggest, it may be possible that epigenetic changes could provide a substrate—or biological foundation—for a fast-track form of natural selection, which enables a species to rapidly adapt to their environment.

One of the most convincing studies of epigenetic inheritance was carried out by Dr. Lars-Olov Bygren, a preventive health specialist at the Karolinska Institute in Stockholm, Sweden.

He looked at human lifespans in Norrbotten, a remote, snow-swept province in the far north of the country. The climate is so harsh there that it causes food harvests to fluctuate from sparse to overflowing—from feast to famine. Historically this meant that the local people's food intake would vary wildly from year to year. The researchers found that a single period of extreme overeating—binge eating—could cause a man's grandsons to die an average of thirty-two years earlier than if the grandfather's food intake had been more moderate. In subsequent research, similar life-shortening effects were found to have an impact on the women in Norrbotten too. Like Suter's work, Bygren's groundbreaking—and disturbing—research suggests that our own eating habits can affect those of our children and grandchildren, and possibly beyond. What we eat today can trigger a biological chain of events that reverberates down through our ancestral line.[38]

But it's not just what we eat. The chemicals and drugs we are exposed to can also have intergenerational effects. A report by Kara Rogers in the January 2012 issue of *Scientific American* suggested that the widespread use of the synthetic estrogen compound diethylstilbestrol (DES) in the mid-twentieth century to prevent miscarriages in pregnant women dramatically increased the risk of birth defects. "It is also associated with an increased risk for vaginal and breast cancers in daughters, and an increased risk of ovarian cancer in maternal granddaughters of women exposed to DES during pregnancy." Dr. Rogers noted that these findings were corroborated in mice studies which suggested that the abnormalities were still present two generations later, and thereby indicating an epigenetic effect.

These examples show how epigenetics can, in certain situations, cause physical changes, such as to a person's weight, color, appearance and health, to be inherited. But what about behavioral and psychological changes? Does epigenetics have the power to affect the human psyche and cognitive abilities? Can it activate our inner geek? Theoretically, yes.

What we do know is that epigenetics plays an important role in neuropsychiatric diseases, such as schizophrenia and bipolar disorder. In one of the largest twin studies performed for any complex disease to date—involving twenty-two identical twin pairs—researchers at the Institute of Psychiatry, King's College London, demonstrated that neuropsychiatric disorders could be expressed differently in a person, even in two people who are genetically identical. Dr. Jonathan Mill, who led the study, believes there may be an epigenetic mechanism at work. "Our findings suggest that it is not only genetic variations that are important. The epigenetic differences we see may tell us more about the causes of schizophrenia and bipolar disorder."

One of Dr. Mill's colleagues, Professor Tim Spector, who is head of the Department of Twin Research at King's College and author of *Identically Different: Why You Can Change Your Genes*, said, "Up to a few years ago I believed genes were the key to the universe. But over the last three years, I have changed my mind."[39] Spector, who in 1993 established the Twins UK registry, one of the largest databases of its kind in the world, cited the example of Nicky and Louise, thirty-year-old identical twins. These women, like most twins, have much in common—they share similar tastes in clothes, food and drink and exercise regularly. They also avoid cigarettes, drugs, gambling and dangerous sports. When it comes to romance and sex, however, they are poles apart. Nicky has had just five men in her life, while Louise has had twenty-five. When they were fifteen years old they learned their father had kept a secret mistress for many years. Each girl reacted differently to this change to their relational environment, and soon developed very different sexual behaviors.

Such divergent outcomes—which feature frequently in Spector's book—have prompted him to question whether a person's genetic inheritance is immutable, and led him to a growing appreciation of the role of epigenetics. "There are now four drugs on the market with epigenetic effects that can switch genes on or off," he said, "and there are forty more in development. . . . We and our genes are more flexible than we thought."

Unfortunately, this flexibility means that we are more likely to pass on the impact of negative experiences to our children. A recent report by a team at the Emory University School of Medicine in the United States showed that mammals (mice) can inherit the phobias and anxieties of their parents. "The experiences of a parent, even before conceiving, markedly influence both structure and function in the nervous system of subsequent generations," the report said.

If epigenetic mechanisms can cause negative conditions in the human brain, such as neuropsychiatric disease (that is, mental illness which has a neurological basis), then it would seem logical that they can work the other way too, and have a beneficial effect. Spector believes this to be the case. When asked by a new grandparent, "What can I do to help my grandchildren epigenetically?" he responded that "There is research that shows that mothers who stroke and show affection to their babies can alter their genes and even the genes of their unborn grandchildren. Cold mothers who deny their children cuddles have the opposite effect on their genes. But if children brought up by cold mothers are transferred at a young age to surrogate mothers who show tenderness towards them the effects on the genes can be reversed—and these changes can last several generations. So, to use the now redundant terms of the old debate: 'nurture' can effectively alter what we thought of as 'nature'—our genes. These studies have so far only taken place with rats, but the evidence suggests that the same effect occurs in humans."

If true, this phenomenon would support Dr. Suter's concept of "biological ancestral responsibility."

Epigenetics may also play a role in transforming a dysfunctional orchid personality into a highly functioning one by, say, affecting the way the CHRM2 gene that is associated with alcoholism and other behavioral conditions is expressed. Also, given what we now know about the potential epigenetic heritability of physical attributes, it is not unreasonable to postulate that changes to our cognitive functions may be inherited through the same process—even if only for a limited time. In fact, given the sensitive and malleable nature of the human brain—which is constantly rewiring itself in response to stimuli—it would seem to be highly susceptible to epigenetic influences.

Indeed, epigenetics may help explain why our cognitive functions appear to be evolving so rapidly in the Anthropocene—faster than natural selection would allow. As any parent (including me) knows, many children and teenagers today are able to operate complex new technology with astonishing dexterity, as if they were born to it. The geek generation is the first in history that is capable of teaching their parents about new technology, rather than the other way around.

They still have the same DNA—or hardware—as their parents, but it appears that their cognitive software has been updated. Perhaps this occurs through the epigenetic process. Joseph Ecker, a Salk Institute biologist and leading epigenetic scientist, put it this way to *Time*'s John Cloud, "I can load Windows, if I want, on my Mac. You're going to have the same chip in there, the same genome, but different software. And the outcome is a different cell type."

It could be argued that what I am describing here is not really evolution because the timeframes are too short; the DNA does not change, and, let's be frank, there are so many unknowns in the process. But if we apply the broader definition of evolution as being "the historical development of a biological group, as a race or species" (*Merriam-Webster*), then it would appear that epigenetics does play a role by speeding up our response time. In the dynamic environment of the Anthropocene, epigenetics is the geek's friend.

Although we have yet to unlock the secrets of epigenetics, there is a growing appreciation of its impact. As Kara Rogers wrote, "[T]hey [the epigenes] do lurk, and silently, they exert their power, modifying DNA and controlling genes, influencing the chaos of nucleic and amino acids. And it is for this reason that many scientists consider the discovery of these entities in the late 20th century as a turning point in our understanding of heredity, as possibly one of the greatest revolutions in modern biology—the rise of epigenetics."[40]

Let's recap. So far we have seen how our world is increasingly driven by human-made (unnatural) influences as we venture further into the Anthropocene. These influences are not just changing the physical landscape; they are also transforming the psychological environment in which humans live. New technologies such as social media are systematically rewiring the neural pathways of our society by changing the way we interact with each other. They are encouraging—demanding—us to think more abstractly and conceptually, and to be adept at multitasking.

This has created a digital greenhouse, which—like Silicon Valley—is nurturing the expression of a different range of cognitive, emotional and psychological traits—many of which were previously considered to be disadvantageous. This greater tolerance for neurodiversity is helping many people to transcend their limitations and even turn their weaknesses into strengths. Such people could be hypersensitive orchids, or those with certain forms of autism, ADHD, dyslexia, or any number of so-called neurological or physical weaknesses. These people now have more scope to exploit their innate talents in an environment that engages directly with their cognitive functions rather than their physical ones. They are less confined by their biological or genetic makeup, and have more potential to self-evolve.

It is also likely that this new digital greenhouse will produce more orchids—and in greater diversity—with each new generation. This is because the Anthropocene is conducive to the genes that underlie the orchid-dandelion phenomenon (sensitivity hypothesis) and it switches them on. As the evolutionary biologist Rob Brooks has explained to me, "When the environment is static, evolution tends to crawl along. But when it changes quite dramatically, as it has during the Anthropocene, evolution tends to leap forward. This is because the new environment changes what it selects for (that is, orchids in this case), and the interaction between the genotype and the environment releases a lot of genetic variation."[41] In other words, the digital greenhouse is fostering new geek varieties to bloom, thereby creating a richer pool of talent. It may also enliven entire cultural groups, as we have seen with the Indian software designers.

This phenomenon may help to explain, for example, why more people are now being diagnosed with some form of autism. According to Benjamin Wallace, writing in *New York* magazine:

"Before 1980, one in 2,000 children was thought to be autistic. By 2007, the Centers for Disease Control was reporting that one in 152 American children had an autism-spectrum disorder. Two years later, the CDC updated the ratio to one in 110. This past March, the CDC revised the number upward again, to one in 88 (one in 54, if you just count boys, who are five times as likely to have one as girls). A South Korean study from last year put the number even higher, at one in 38. And in New Jersey, according to the latest numbers, an improbable one in 29 boys is on the spectrum."[42]

Many of these increases are likely to be the result of greater public awareness, broader definitions of autism, and better incident reporting. But these explanations may not account for all the rises. According to Mark Roithmayr, president of Autism

Speaks, an autism research and advocacy group, "Only part of the increase can be explained by better and broader diagnoses." He told *Time*, "There is a great unknown. Something is going on here, and we don't know what."[43] Some scientists suspect that this unknown factor may be environmental. Dr. Thomas Insel, Director of the National Institute of Mental Health, wrote an article in April 2012 titled "The New Genetics of Autism: Why environment matters," in which he suggested that "environmental factors can cause changes in our DNA that can raise the risk for autism and other disorders."[44]

Indeed, the Anthropocene may actually be selecting for certain autism traits that confer cognitive advantages—and other orchid behaviors—in order to cope with the increased technical complexity of the modern world. A vocal proponent of this view is Juan Enriquez, a futurist at Harvard Business School, who suggests rising autism rates are an adaptive response to a data-intensive environment. "When you think of how much data is coming into our brains, we're trying to take in as much data in a day as people used to take in, in a lifetime." Rising levels of autism are a natural response to this challenge, he says, and are helping to lay the foundation for "the next evolutionary step."[45]

This process may be accelerated by "assortative mating," whereby people seek to bond with those whose characteristics resemble their own, thereby encouraging successful non-neurotypical people with, say, Asperger's, to mate and produce offspring. Whereas in previous generations "those with Asperger's syndrome tended to be loners," as Steve Silberman explained. "They were the strange uncle who droned on in a tuneless voice, tending his private logs of baseball statistics or military arcana; the cousin who never married, celibate by choice, fussy about the arrangement of her things, who spoke in a lexicon mined from reading dictionaries cover to cover."[46] Now, of course, given the right environment (digital green-house), such a person could end up on the cover of *Fortune*.

We have also looked at how our minds are highly responsive to external conditions and why intelligence (or IQ) is not as fixed as previously thought. Our brains are actually highly malleable and are constantly evolving in response to the stimulation they receive. David Shenk (author of *The Genius in All of Us*) suggests that it is helpful to think of our genome as "a giant mixing board with thousands of knobs and switches. Genes are always getting turned on/off/up/down by hormones, nutrients, etc."

This is precisely what the Anthropocene is doing—switching knobs on and off, activating others, and rewiring our neural processes—so that we become better adapted to the pressures of our techno-centric age. It is also possible that this process is being accelerated by epigenetics, which influences the way our genes are expressed in response to the environment, and may even transmit them to succeeding generations through soft inheritance—thereby speeding up the collective learning process.

The cumulative impact of all these developments—technological, genetic and cultural—is to create a cognitive revolution in our society. The Anthropocene is pushing us to become smarter so that we can better cope with the pressures of the technological age, thus creating an evolutionary virtuous circle.

At the forefront of this revolution are the geeks, whose unique traits and characteristics enable them to resonate with the digital nervous system of our high-tech era—as demonstrated by the extraordinary success many of them have achieved in recent years. These people don't just engage passively with the new environment—as consumers of digital delights—they engage with it proactively as creators of a brave new world. They are the personification of the Anthropocene and are intent on remaking it in their own image. They have the intelligence, capabilities and resources to do so. The more successful and influential they become, the more others emulate them, thus creating a geek culture—or meme—that feeds on itself and further accelerates the cognitive revolution.

Let's now take a closer look at these geeks, who seem destined to inherit the Earth.

# Chapter Three

# *The Rise of the Geek*

## *What makes a geek?*

During the nineteenth century in America, a "geek" was a carnival performer or wild man who would bite the head off a live chicken or snake—and swallow it. His job was to warm up the audience for the main show that followed. It drew on a similar act performed a century earlier by circuses in Austria-Hungary, where people called *gecken*—or "freaks"—were used as attention-getters. The word survives today in the Dutch language as *gecken* meaning "crazy."

It wasn't until 1952, with the publication of Robert Heinlein's short story "The Year of the Jackpot," that the word "geek" became associated with technology. This curious tale features a middle-aged man named Potiphar Breen, who is a statistician working for insurance companies, and who buries himself in journals about astrophysics during his spare time. At one point during the story, he is concerned about the fate of a Russian cosmologist who has written about the stability of G-type stars, and appears to have been liquidated by the authorities.

"The poor geek," Breen laments. And in uttering those words he gives birth to a new genre of character—the technocentric being.

Over the years, the term "geek" has evolved considerably, and is now used to describe a person who is very knowledgeable about a particular subject. Such people tend to be intellectual, intense and passionate about their field of interest—so much so that their behavior often verges on the obsessive. They may not be the snake-biting madmen of old—although some may disagree with this—but they do tend to be non-neurotypical in a psychological sense.

Most geeks share a deep connection with the digital-scape of the Anthropocene, and spend much of their time online in pursuit of their interests. Within their virtual cocoon they can become cut off from their physical-world lives, much to the dismay of their family and friends. The novelist Julie Smith once described a geek as:

"[A] bright young man turned inward, poorly socialized, who felt so little kinship with his own planet that he routinely travelled to the ones invented by his favorite authors, who thought of that secret, dreamy place his computer took him to as cyberspace—somewhere exciting, a place more real than his own life, a land he could conquer, not a drab teenager's room in his parents' house."[1]

This is a harsh characterization, but it does highlight an almost universal truth about geeks—they seek a world they can control—even if it is in their own imagination. This is what distinguishes them from their cerebral brethren such as nerds, dorks, dweebs, noobs, gimps, spods and gumps. A geek doesn't seek knowledge for knowledge's sake—like a nerd does—he wants to actually apply this knowledge to make changes, to bring forth "a land he can conquer," as Smith wrote. This bias for action is what makes geeks so powerful in the world. As Richard Clarke once explained during an interview on *The*

*Colbert Report*, the difference between a nerd and a geek is that "geeks get it done."

For this reason, geeks are often associated with pragmatic, real-world disciplines, which define their identities. So we have, for example, engineering geeks, mathematics geeks, computer geeks, science geeks, financial geeks, astrophysics geeks, entrepreneurial geeks, technology geeks and so on. They like to apply themselves, not just cogitate. So a mathematics geek, for instance, might use a multivariate calculus to determine how to optimize the dimensions of a pan to bake a loaf of bread. Or an astrophysicist geek may be inclined to ditch the sat-nav and plot her course by the stars—just for the heck of it—and because she can.

The other thing about geeks is that they tend to exhibit an almost childlike curiosity and playfulness, no matter how old they are. Like a bespectacled version of Peter Pan they never lose their sense of wonder. This characteristic of retaining childlike characteristics in a mature member of a species is known as neoteny. Unlike other species, which lose their playfulness on maturity, most humans retain the capacity for childlike behavior as they get older. But geeks seem to have an extra-strength version of neoteny, which endows them with an eternal sense of starry-eyed wonder and playfulness, and helps them to remain more flexible and receptive to new information.

You can often see this trait in middle-aged scientists and technologists who are barely able to contain their enthusiasm for their subject. This behavior was on graphic display during the announcement of the discovery of the Higgs Boson particle in Lucerne, Switzerland, on July 4, 2012, when an entire auditorium of scientific geeks—many of them quite advanced in years—erupted in squeals of childlike delight. You would never see a room full of bankers act this way, no matter how high the interest rates soared. Geeks also have a habit of redefining language on their own terms, particularly in their use of superlatives. One of Steve Jobs's favorite expressions, for example, was "insanely great."

The phenomenal success of people like Steve Jobs, Bill Gates, Mark Zuckerberg and others has enabled geeks everywhere to become more self-confident about their identity and status in society. They no longer hide their geekiness. Indeed, for many of them it is a badge of honor, as is reflected by events like Geek Pride Day, which is held in Spain on May 25 each year, and the numerous geek-related sites that have sprung up on the web, where a Google search of the word "geek" will reveal about 220,000 results.[2] They also appear in their own television shows such as *Big Bang Theory*, which tracks the chaotically humorous lives of some intellectual twenty-somethings. There are also geek dating sites for people who place a high value on cerebral intercourse. Geekness is no longer a predominantly male domain either, as more females gravitate toward high-tech jobs and adopt geek habits and lifestyles.

The fashion industry, too, has embraced the geek look. When Russell Westbrook, the American basketball superstar, started turning up at press conferences wearing thick spectacles with horn-rimmed frames without lenses, it caused quite a stir in NBA (National Basketball Association) circles. Westbrook had long been known for his wildly colored and patterned shirts, but the geek glasses were a step in a different direction.

The rise of "geek chic"—as exemplified by the glasses— also signals a shift in young people's attitudes to education and learning. NBA star Dwayne Wade told American Public Media's Marketplace, "the glasses end up sending the message that 'it's cool to be smart, it's cool to be educated.'" Indeed, in US high schools, the sports "jock" is no longer the top of the pecking order—the geek is. A recent survey conducted by Modis recruitment agency showed that a clear majority of young Americans say they would rather be called a geek than a jock, because they are seen as more intelligent. Amanda David, nineteen, who is studying management and architecture at MIT (a cerebral hot spot), was quoted in *USA Today* by journalist Haya El Nasser saying students are now much more well-rounded. "Now, at MIT, the captain of the crew team might be the best at programming," she said, "the stereotype is certainly being challenged . . . Students can make all these things that society values. That, right there, is making the typical geek cool."[3]

Sherry Turkle, an MIT psychologist and author of *Alone Together: Why We Expect More from Technology and Less from Each Other*, also believes there has been a fundamental change in the way geeks are perceived in recent years. "We're all techies now. We're dependent on these people, so there's a power shift, a new kind of respect," she told El Nasser. Turkle cited the example of the Genius Bars at the Apple stores, where people come to have their technical problems solved by the geeks who work there. Customers are often desperate to have their email or Facebook working again. "They're close to hysterical," she said. "They're quite dependent on these geniuses to help them, and those geniuses do. Those young men and women [who were once derided as nerds] are not objects of derision."

Turkle's view is shared by Jack Cullen, President of Modis, an IT recruitment company, who said, "Being a geek has gone mainstream. It may be because of Americans' increasing dependence on, and comfort with, technology. Or the prevalent images of former geeks who now successfully lead multi-billion dollar technology companies."[4]

A telling example of this reputational transformation can be seen in the recent James Bond film, *Skyfall*, in which the character named "Q"—who provides 007 with his gadgets—is no longer a geek to be ridiculed. He is someone to be reckoned with, and in many ways, is a match for Bond himself. As the *Guardian*'s James Ball wrote, "For decades, Q represented the most traditional British form of geekdom: the boffin—a middle-aged man obsessed with slightly naff gadgets, working from the hi-tech equivalent of a shed, an exasperated comic foil to our suave protagonist." But now Q has returned in the latest Bond film as, "the kind of geek Apple design supremo Jonathan Ive would create: a Zen-like twenty-something in an on-trend chunky knit cardigan and fashionable specs. Gone are the naff gadgets, in favour of a couple of immaculately designed tools (and an incidental cyberwar). Most thrillingly for geeks, the modern Q can hold his own when verbally sparring with our action hero—a delightful payoff for millions who dreamed of exactly that during tougher times at school and beyond. Geeks have hit the cultural, technological and economic mainstream. Heck, just take a look at real-life MI6 job adverts (they exist) and you'll see far more looking for CVs like Q's than Bond's—the main challenge is simply whether they can pay enough."[5]

Ball noted that it is a good time to be a geek, but stressed that we should not confuse the image with the substance. "The real shibboleth of geekdom," he wrote, "is a deep, unabashed fascination with either one subject area or many: an obsession with detail, with exploration, and (frankly) with learning. This tends to trend in areas where there's a lot to learn, a lot to discuss, and a lot of ways to show off."

Being perceived as a geek can also significantly enhance a person's employment prospects, which is why many people choose to wear glasses for job interviews. A study by the UK College of Optometrists in 2011—perhaps unsurprisingly—found that 40 percent of those with 20–20 vision would consider wearing clear lenses if it would improve their chances of getting a job.[6] Another 6 percent would put on glasses to feel fashionable, and 9 percent think spectacles can make them look more attractive. Psychology professor Cary Cooper, from Lancaster University, said, "It is not surprising that businesses want to employ intelligent staff, but the idea that intelligent people wear glasses is an old stereotype that has not gone away."[7] The stereotype that has changed, however, is that of intelligent people being boring or less attractive than their more social or sporty peers. They are now seen as cool. They may wear these big dorky glasses and not be good at making eye contact, but as Robert Thompson, Trustee Professor of Popular Culture at Syracuse University, quipped, "If you're worth $20 million, who needs eye contact?"[8]

Geeks have certainly come a long way since the bad days of the *gecken*. They are no longer the warm-up act, they are the main act, and increasingly want to run things their way. Balaji Srinivasan, a Stanford University lecturer, told an audience of aspiring tech entrepreneurs in October 2013 that Silicon Valley should secede from the United States, and operate under its own laissez-faire rules. "The best part is this," he said, "The people who think this is weird, the people who sneer at the frontier, who hate technology—they won't follow you out there."

The sheer audacity of the techno-elites has encouraged many non-geeks to cling to—and promote—the old geek stereotype of benign quirkiness. It makes them seem less threatening. It's like calling a pit-bull terrier a poodle, or Mike Tyson, Maryanne. So no matter how rich or influential or powerful a geek becomes, it's okay, because they are still just a geek at heart. It also helps that many geeks appear to be non-neurotypical, and may, for example, exhibit signs of ADHD or being on the autism spectrum.

"To some degree," wrote Benjamin Wallace in *New York*, "the [autism] spectrum is our way of making sense of an upended social topography, a buckled landscape where nerd titans hold the high ground once occupied by square-jawed captains of industry, a befuddling digital world overrun with trolls and avatars and social-media 'rock stars' who are nothing like actual rock stars. It is, as the amateur presidential shrinks would have it, a handy phrase for the distant, cerebral men with the ambition and self-possession necessary to mount a serious run for the White House. When quants (financial geeks) and engineers are ascendant, when algorithms trump the liberal arts, when Kim Kardashian and Justin Bieber tweet about the death of Steve Jobs, when the hyper-specialist has displaced the generalist and everyone is *Matrix*-ed into the Internet, it's an Other-deriding tool to soothe our cultural anxiety about the ongoing power shift from humanists to technologists. As the coders inherit the Earth, saying someone's on the [autism] spectrum is how English majors make themselves feel better."9

In effect, the term "geek"—and its connotations of non-neurotypical traits such as ADHD and autism—provides linguistic reassurance to non-geeks, who perhaps worry that they don't have what it takes to compete on their terms.

It would be a mistake, however, to assume all geeks are the same. Although they do have much in common, there are fundamental differences in terms of their motivations, aspirations and how they relate to the Anthropocene. Like the orchids they often appear to emulate, geeks have evolved into a variety of different species—all of them fascinating, most of them brilliant, and some of them just plain scary. We have barely begun to comprehend their impact on society. One thing is certain, though; this innately disruptive force is revolutionizing our world—for better or worse.

Over the next pages, we will examine the geek archetypes in more detail.

## *The purist* (Geekus purizenicus)

The purist geek is the most single-minded, obsessive and ruthless geek of all. He or she will dedicate years, even decades, in pursuit of a singular vision. They will let nothing stand in their way, and will endure extraordinary hardship and privation to achieve their goal. Sometimes their personal zeal is so intense that it can take on an almost religious quality.

Indeed, in another life a purist geek could have been a Zen Buddhist monk—the sort of person who would meditate for hours on bare floors, in freezing winters with little food, while focusing their steely concentration on a single flickering candle. The monk would put up with anything in order to attain spiritual enlightenment, or *satori*. One of the best descriptions of this mindset can be found in a little book called *The Method of Zen* by Professor Eugen Herrigel, a leading Western scholar on Buddhism in the twentieth century. He wrote that life in the austere Zen monastery is all about "concentration in order to meditate—it requires the capacity to concentrate on the same

thing for hours, days and weeks on end. This presupposes firm-
ness and steadiness of will and no less than a clear intellect."

In 1975, an intense young man joined the Tassajara Zen
Mountain Center in San Francisco, the first Buddhist monastery to be built outside Asia. The man had just returned from
a trip to India where, with a college friend, he had travelled to
the ashrams of Neem Karoli Baba and Hariakhan Baba, and
explored the northern states of Uttar Pradesh and Himachal
Pradesh, and the great sprawling metropolis of Delhi. By the
time he returned home he was so deeply inspired by what he
had experienced that he was barely recognizable—his head was
shaved, he wore traditional Indian clothing and began experi-
menting with hallucinogenic drugs such as LSD.

His name was Steve Jobs.

Like many people of his generation, Jobs was swept up
in the counterculture movement of the '70s, which had begun
years earlier as a rebellion against materialistic and capitalist
values, and been fuelled by anti-Vietnam War sentiment. This
strident protest era eventually gave way to a more moderate
period in which young people—particularly hippies—became
interested in spiritual enlightenment. In their quest to live
more fulfilling lives, many of them turned to Eastern religions,
including Buddhism. They sought guidance in the writings of
mystic luminaries like Alan Watts, an exponent of Zen Buddhism; Alan Ginsberg, the iconoclastic poet; and the cult-like
figure of Jack Kerouac, who wrote *The Dharma Bums*, an ode
to the simple life and Buddhist principles.

The epicenter of this counterculture revolution was San
Francisco—the "flower power" city—which had become a
spiritual crossroads between East and West, and a popular des-
tination for those seeking salvation—or at least to be pointed

in the right direction. It was the perfect place for Jobs to be. *Wired*'s Steve Silberman described how Jobs would drive along the twisting dirt road in the mountains above Carmel, and soon arrive at the entrance to the Tassajara Zen Mountain Center.[10] Here, he would take off his shoes, prop himself cross-legged on a cushion, and sit for hours on end "facing the wall," and observing the activity of his own mind. Sometimes he would spend weeks here, under the supervision of the monastery's legendary founder, Shunryu Suzuki, and his mindful monks. At one point, Jobs took his meditation so seriously that he considered taking up residence at the renowned thirteenth-century Eihei-ji monastery in Japan—a rigorous boarding school for hopeful Buddhist illuminati.

This period of intense self-reflection reinforced Jobs's commitment to the Zen values of simplicity, clarity and focus, and how they could inform his life and work. So much so that when he founded Apple Computers in 1976, these purist values would become the DNA of his company and eventually permeate every aspect of its operations.

There was one value in particular that Jobs—and Apple—lived by. Focus. While other companies, such as Sony and Nokia, relentlessly churned out more and more products to satisfy every conceivable consumer whim, Apple made just a few. Like a Zen-trained Japanese sword-maker, Apple concentrated on refining and perfecting the work at hand. It resisted the temptation to add more embellishments, gimmicks and complexity. For Jobs and his team, mastery lay in stripping away the unessential to reveal what was valuable and useful.[11] In an interview in *Fortune* in 2008, Jobs explained, "Certainly the great consumer electronics companies of the past had thousands of products. We tend to focus much more. People think focus means saying yes to the thing you've got to focus on. But that's not what it means at all. It means saying no to the hundred other good ideas that there are. You have to pick carefully."[12]

Jobs's obsession with focus and simplicity also informed other aspects of his life, including his fashion sense—black turtleneck skivvy—and his living space. When he purchased a sprawling fourteen-bedroom Spanish colonial mansion in Woodside, California, in 1984, he kept it almost unfurnished— monastery-like—for nearly a decade, except for a music system, a 1966 BMW R60/2 motorcycle in the living room, and some rudimentary furnishings. Within this uncluttered space, he could think and ponder the future, free of distraction. Similarly, when Apple launched its phenomenally successful retail stores at the turn of the century, they were designed as monastic halls of light and space, linked by glass staircases—places where customers could contemplate on the wonders of Apple's products. Indeed, for many avid customers, Apple products have appeared to be endowed with magical, almost supernatural qualities. When the iPhone was launched in 2007 it was dubbed by many Apple enthusiasts as the "Jesus Phone" because of its ability to perform so many miraculous functions. And of course, there has been the iPod, iMac, iTunes and the iPad. In fact, by the time Jobs prematurely passed away in 2011 from cancer, he had transformed six different industries—computing, phones, music, film, retail and photography.

For his achievements, Jobs has been described as legendary, visionary, a design perfectionist, and the Father of the Digital Revolution. He also became fabulously wealthy—worth nearly US$7 billion. In January 2012, when young adults aged sixteen to twenty-five were asked to identify the greatest innovator of all time, Steve Jobs came second behind Thomas Edison.

In 2012, the global business magazine *Fortune* named Steve Jobs the "greatest entrepreneur of our time," explaining that "though he could be abusive and mean-spirited to people who threw themselves into their work on his behalf, Steve Jobs has been our generation's quintessential entrepreneur. Visionary. Inspiring. Brilliant. Mercurial."[13] At one point, *The Economist* ran a cover picture of him looking like a messenger from heaven, holding an iPhone-like stone tablet, as if it conveyed the words of God.

Like any deity, however, Jobs was not an easy person to work with. The Buddha had advised his disciples "to discover your work, and then give yourself to it with all your heart." Jobs certainly did that. But in doing so, he could be so demanding, and was such an obsessive perfectionist, that it would drive many coworkers around the bend. His managerial style was, in many ways, the opposite of what was expected of the modern civilized manager. He could be highly autocratic, secretive, maniacal, and prone to outbursts of aggressive and sometimes vindictive behavior. During a tantrum his criticisms could wound deeply. After the debacle of Apple's Mobile Me launch—a rare failure—Jobs told the team, "You should hate each other for having let each other down."

He was also the consummate micromanager, and had his fingerprints over every design that left an Apple factory. Steve Silberman described Jobs as a "Zen fussbudget," who "paid precise, meticulous, uncompromising attention to every aspect of the user experience of Apple's products—from the design of the fonts and icons in the operating system, to the metals used to cast the cases, to the colors on the boxes and in the magazine ads, to the rhyming proportions in the layout of Apple stores."

This attention to detail had been encouraged by Jobs's adoptive father, Paul Jobs, a skilled mechanic, who had advised his son, "You've got to make the back of the fence that nobody will see just as good-looking as the front of the fence. Even though nobody will see it, you will know, and that will show that you're dedicated to making something perfect."

Jobs's uncompromising way of doing things, and his unerring faith in his own judgment, rankled with even the corporate renegades of the computer world. *Fortune* noted that he was "considered one of Silicon Valley's leading egomaniacs." Jeff Raskin, a former colleague, once suggested that Jobs "would have made an excellent king of France."[14]

Barely a year after his death, in August 2012, *Wired* challenged the managerial legacy of Steve Jobs and whether—despite his success—people should try to emulate him. Under the headline Do You Really Want to Be Like Steve Jobs? the cover story listed a range of Jobs's contradictions: "He was a Buddhist . . . and a tyrant; he was a genius . . . and a jerk; His life story has become an inspiration for some . . . and a cautionary tale for others."

The Australian commentator and broadcaster Phillip Adams observed that Walter Isaacson's biography of Steve Jobs revealed him as "a tyrannical, tear-shedding, tantrum-tossing, cold hearted bully. And this, let me remind you, is the official biography."[15]

Given the idiosyncrasies of Jobs's managerial style, how did he become so phenomenally successful? Why wasn't he undone by the foibles of his personality and behavior, given that they so blatantly contradicted those espoused by nearly every modern business school? How did he succeed in spite of himself?

## Zeitgeist and zazen

He did so for the same reason Mark Zuckerberg succeeded, and countless other successful geeks today do. That is, because the new environment of the Anthropocene tolerates—and actually encourages—a far greater range of behavioral and cognitive diversity than previous times in human history. So even an extreme "purist" like Steve Jobs—who might have been crucified in an earlier time for alienating too many people while trying to build his company—could now be empowered to act without sanction. The disruptive technologies of the Anthropocene, particularly the Internet, have leveled the playing field—sociologically, cognitively and culturally. Today a CEO of the most profitable corporation on Earth—which Apple was in late 2011—is not required to conform to the old stereotype of the solid captain of industry. As long as they *perform*, is all that shareholders really care about today.

Importantly, because geeks tend to care less about what other people think, they are less constrained by orthodox opinion. Jobs, for example, loathed consumer focus groups and had little requirement for intermediaries and consultancies to tell him what was going on. He preferred to tap into his intuition—and the technological zeitgeist—to determine what needed to be done. This fiercely independent mindset meant that Jobs was less encumbered by old-fashioned notions of what a computer, or phone, or music player or tablet or animated film should be. He could envisage them afresh. Jobs once told the 2007 Macworld Conference and Expo, "There's an old Wayne Gretzky [the great ice hockey player] quote that I love, 'I skate to where the puck is going to be, not where it has been.' And we've always tried to do that at Apple since the very beginning."

Jobs's vision of intuitive computing became so influential that it exerted an almost hypnotic effect over the industry, and became known as Jobs's "reality distortion field." People exposed to it, whether they were customers, suppliers, Apple employees, or journalists, would often sublimate their own perspective in favor of Jobs's vision. Critics suggested that this created unrealistic expectations that would come back to bite Apple. Sometimes they did, but not often.

What many critics might not have appreciated is that in the digital age of the Anthropocene, unlike in decades earlier, Jobs was able to create his own version of reality—on his own terms. He resonated so strongly with the technological zeitgeist that he had become one with it, and to a very real extent was a cocreator of it. This endowed him with the power to sometimes bend perceptions to his will.

Over time, Steve Jobs the person, and his Apple products, began to converge into a corporate singularity—a trinity of man, machine and environment. Everything he created—the hardware, software, music, films, apps and retail stores—became inextricably linked to form an entire digital ecosystem, which in turn, became a cornerstone of the Anthropocene. Jobs's most enduring legacy, however, may not be his products or software—but an attitude of mind—which he articulated in a commencement speech to Stanford University graduates in 2005—two years after he had been diagnosed with pancreatic cancer. "Your time is limited," he urged, "so don't waste it living someone else's life. Don't be trapped by dogma—which is living with the results of other people's thinking. Don't let the noise of others' opinions drown out your own inner voice. And most important, have the courage to follow your heart and intuition. They somehow already know what you truly want to become. Everything else is secondary."[16] The call to "follow one's heart" is not a new message, of course. But Jobs's words may have a particular resonance for the types of people discussed earlier in this book who suffer an innate affliction such as ADHD or Asperger's syndrome—and feel pressured to conform to a social "norm." Such people may be inspired by Jobs's call to resist these pressures and to celebrate and capitalize on their Neurodiversity—and to follow it wherever it may lead them.

## *The financial geek* (Gordonus gekkonus)

In the James Bond film *Never Say Never Again*, Bond (Sean Connery) plays a 3-D video game called Domination against his nemesis Maximillian Largo (played by Klaus Maria Brandauer). Each time a player loses a point, he is given an electric shock until the pain becomes too great to bear. Bond eventually wins, of course, but not before both men have experienced excruciating agony. This is a game of severe consequences.

All around the world today, on every major bank's trading floor, there are testosterone-fuelled young men playing a financial version of Domination. They sit before big screens that gush torrents of data, which they deftly sift through, while making calculated guesses. Then they make their move—betting millions, or even billions, of dollars in a few keystrokes—on stocks, options, derivatives, currencies, commodities, default swaps—anything that moves up or down in the markets. It is a high-stakes game and not for the fainthearted.

Financial traders are the 007s of the banking world. They may not look like James Bond, but many of them can afford to emulate his lifestyle of fast cars, expensive champagne, Italian suits, and, yes, beautiful women. Unlike Bond, however, when a trader comes face to face with his nemesis—a failed trade—they rarely suffer the consequences. For they have bet other people's money—people like retirees, mums and dads, and employees—whose pension funds have been entrusted to their employer. It's the trader's job to maximize these clients' returns, for which they are paid handsomely—sometimes earning massive bonuses. Many top traders operate in the City of London, which reverberates with so much mercantile activity that it ranks as one of the financial capitals of the world. The "City" is the Anthropocene on steroids. London has always had a reputation for producing "wheeler-dealer" personalities, from the fast-talking barrow boys of the East End, to the buccaneering capitalists of the eighteenth and nineteenth centuries, who expanded their empires under the protective sails of the Royal Navy.

But the modern financial trader is no barrow boy or buccaneer; he is essentially a geek—a high-caliber video gamer who is exceptionally good at interpreting data and making lightning-quick decisions. These traders are invariably young, ambitious, tech-savvy and usually male. Given the risks involved, they also have nerves—or rather, balls—of steel. Think of the trader geek as an upgraded version of Gordon Gekko (from the film *Wall Street*), with an Intel computer chip implanted in his cerebral cortex. His career will not be long-lived, because in the fast and furious world of trading, many burn out by the time they are thirty-five years old. It is no surprise, therefore, that trader geeks tend to be impatient. According to Andrew Haldane, Executive Director for Financial Stability at the Bank of England, "Most traders' brains harbor the impatience gene."[17] Delayed gratification is not their strong point.

A few floors away from the hustle and bustle of the trading floors, you will find another species of financial geek hard at work. These are more cerebral and low key than their testosterone-fuelled brethren, and operate in an arcane world of numbers and statistics. Such people are known as "quants." It's their job to create computer-simulated models that predict how the markets will operate in order to help bankers and traders to make better decisions. The quants are usually brilliant PhD-wielding graduates drawn from top universities such as MIT in the United States and Cambridge University in the United Kingdom, who can apply their quantitative magic to simulate practically any scenario in the real world. But whereas once they spent their time deciphering string theory in the hope of winning a Nobel Prize, now they look for ways to make money in an industry where multimillion-dollar bonuses are not uncommon. They have struck a Faustian bargain with Wall Street or the City of London.

To create their models, the quant geeks employ artificial intelligence and concepts like "stochastic volatility" and "the super statistics of labor productivity" to penetrate the mysteries of market behavior. The models enable bankers to conduct business at lightning speed and in staggeringly large volumes. Whereas up until a few decades ago traders had to laboriously and meticulously analyze figures before completing a transaction, now they can act almost instantaneously. The models also underpin the automated high-frequency trading programs that now account for over 60 percent of all equities trades.

I first came across a quant in the early 1990s when I began work at a financial services company in Sydney. A colleague was showing me the ropes and pointed to a tall young man who suddenly appeared from nowhere at the other side of the room. He had straggly hair and was dressed in a long black coat.

"That's Darth Vader," my handler said. "You won't see much of him because he is locked away most of the time. And we don't even know his real name."

"What does he do?" I asked.

"He designs 'black boxes' for the investment-management guys, which are computer programs that trade stocks and bonds automatically. Experimental stuff."

"Why are they called black boxes?" I enquired.

"Because no one understands them except guys like Darth."

"Must be smart."

"You bet, he's probably the smartest person here—even smarter than the actuaries who consider themselves a cut above the rest of us."

"Then why doesn't he run the place?"

"Because people like him are too dangerous to let loose on the system. You never know where their weird ideas could lead us."

A decade later, when I moved to London to work for a big Swiss-based investment bank, I came across a lot of people like Darth. They were no longer locked away, but roaming freely among us mere mortals. The financial industry's growing dependence on computer models had propelled these geeks to the top of the financial pecking order. Like in other industries, they had emerged from the backrooms and basement research departments, to become the driving force of their companies. Their tech-savvy activities often now account for the major portion of most banks' revenue and generate most of the profits.

To get a clearer sense of why these financial geeks have become so essential to the global financial system, let's take a closer look at one of their key activities: derivatives trading.

## Masters of the black hole

Derivatives—as the name suggests—derive their value from other things. They don't have an intrinsic value of their own but offer investors a way to buy or sell something, such as stocks, bonds, currencies, commodities or interest rates, at an agreed price at a later date. The defining feature of derivatives is that they are one step (or many steps) removed from the real, tangible asset on which their value is based. For example, a home-insurance policy is a form of derivative.

Linda Davies in her book *Into the Fire* compared the job of a derivatives trader to that of "a bookie once removed, taking bets on people making bets."[18]

Investors are keen on derivatives because a derivative can be more flexible than a stock and generate high profits. It is also easier in many markets to borrow money to purchase a derivative than, say, a stock. So you can make bigger bets on derivatives without putting up much collateral, and sometimes, without putting up any at all.

Derivatives have become so attractive to investors that over the past twenty years, more money has poured into derivatives than into all other investment vehicles combined. According to the Bank of International Settlements, by 2011 the total amount of derivatives outstanding was $750 trillion, which is three quarters of a quadrillion.[19] A quadrillion is a number that is normally used only by astronomers when calculating, say, the number of stars in a galaxy, or by people in universities with lots of chalk on their coats. It is 1,000,000,000,000,000 — a mind-bogglingly large number. By comparison, the entire annual economic output of the world (GDP) is about $50 trillion.

Given the sheer scale of the numbers involved, derivatives are sometimes called the "black holes" of the financial world. They are so powerful that they tend to suck many investors into them, and are virtually incomprehensible. Indeed, an investor would in theory need to read as many as 1.125 billion pages to understand the components in the type of derivatives-based security known as a collateralized debt obligation squared, or $CDO_2$, which could contain portions of nearly a hundred million mortgages. This complexity has led many analysts to believe that derivatives are, in effect, "financial weapons of mass destruction."

For financial geeks, however, derivatives are a godsend, because the only way that banks can manage and calculate the value of derivatives is by employing mathematicians and physicists to make sense of them. These specialized analytical skills are so highly regarded that in 1997 the Royal Swedish Academy of Sciences awarded the Nobel Prize in Economics to Professor Robert C. Merton of Harvard University and Professor Myron S. Scholes of Stanford University, to honor the method they jointly developed to calculate the value of options derivatives.

Unfortunately, one year later, a derivatives-based hedge fund in which Merton and Scholes were principal shareholders, Long Term Capital Management, had to be rescued at a cost of $3.5 billion to the taxpayer when the Federal Reserve Bank decided that its collapse would pose a serious threat to the stability of the American markets.[20] Almost ten years after this, in

2007, one of Wall Street's most blue-blood firms, Bear Stearns, was brought down by complex derivatives trading in hedge funds—making it the first major victim of the global financial crisis. Shortly thereafter, Goldman Sachs also revealed it had lost billions through hedge-fund derivatives trades. By the time the financial crisis reached its peak in 2010, it was clear that derivatives trading was responsible for losses totaling hundreds of billions of dollars.

Emanuel Derman, a physicist-turned-financier who had played a central role in the early development of models for derivatives, once warned his employer, Goldman Sachs, of the limitations of derivatives models. He likened their relationship with reality to that between a child's toy car and an actual automobile. More recently, Derman, who is now a professor at Columbia University and the author of *My Life as a Quant: Reflections on Physics and Finance*, wrote, "It's difficult or well-nigh impossible to systematically predict what's going to happen. You may think you know you're in a bubble, but you still can't tell whether things are going up or down the next day." He also cheekily coauthored a financial modeler's version of the Hippocratic Oath, which pledges, among other things, that "I will remember that I didn't make the world, and it doesn't satisfy my equations."[21]

Bright people like Emanuel Derman who challenge the status quo, however, are the exception. For it doesn't matter how sophisticated or intelligent a person (or organization) is, they can be just as vulnerable to the influence of destructive models as a less intelligent person, and in some cases more so. This is because, from a psychological perspective, intelligent people are often more comfortable dealing with synthetic simulations of reality which may be intellectually sound but not practical in the real world. Albert Einstein once remarked on this tendency in a letter to Sigmund Freud, on July 30, 1932: "Experience proves that it is rather the so-called 'intelligentsia' that is most apt to yield to these disastrous collective suggestions, since the intellectual has no direct contact with life in the raw but encounters it in its easiest, synthetic form."[22]

Which brings us to the Achilles heel of the financial geek.

## Skimming not diving

We have seen in this book how the geeks have a special relationship with the information-rich environment of the Anthropocene. They resonate with it. This gives them an edge when it comes to navigating, and capitalizing on, the huge volumes of data that engulf us. Indeed, by 2012 the average person in developed economies was exposed to three times more information per day than they were in the 1980s. This is hardly surprising considering the explosive growth in Internet-based technologies and social media such as Facebook and Twitter. But information is not knowledge. People don't automatically become smarter by being immersed in a sea of data any more than security guards in art galleries become art experts through a process of osmosis. Information must be chewed over, tested and digested before it can become knowledge. Too much information can be a bad thing. This is because the only way that most of us can cope with vast oceans of data is to skim the surface and glean the fragments of information that seem most relevant.

There are certain professions that breed "skimmers": people whose job it is to scan and process information at lightning speed. You have probably guessed whom I am talking about. Financial geeks.

Such people spend their days amid Bloomberg (financial news) screens, sifting through vast data flows, in order to optimize their trades or to perfect their models. In theory, no professional on Earth is better at dealing with the challenges of the information age and making sense of it. After all, that's what their job is.

Yet the financial geeks were among the first to be drowned by the tsunami of the global financial crisis. Why? Because they couldn't detect the tectonic shifts occurring on the ocean floor that would generate the destructive wave. They were skimmers not divers. And they remain so today. Which explains why—years after the financial crisis—we continue to be shocked by periodic massive trading losses at big banks, despite their having been bailed out previously, and despite the enormous efforts they have made to improve their risk-management systems.

The point is that the problem has never been about risk management, per se; it is about information overload. And despite their intellectual prowess, geeks are just as vulnerable to overload as the next person—probably more so—because they tend to have an insatiable appetite for data. They may also ignore warnings from the "real world" because, as Einstein would say, they prefer their "synthetic" versions of reality. Their overreliance on complex financial models enables them to "be wrong with infinite precision." More problematically, the geek's consummate faith in the power of logic makes them susceptible to the notion that the solution to any problem is more information.

In the 1960s an interesting experiment was conducted that demonstrated the folly of having too much information. Two groups of people were shown a fuzzy and indistinct outline of a fire hydrant. The resolution was gradually increased for one group, through a series of ten steps. For the other group, however, the resolution was increased over just five steps. Then the process was stopped at a point where each group was looking at an identical picture and the subjects were asked what they could see. It turned out that the members of the group that saw fewer intermediate steps recognized the picture earlier than the group presented with more steps. The extra information encouraged the latter group to speculate more about what the image was, clouding their judgment. The first group saw the fire hydrant more directly for what it was, unhindered by too many layers of information.

The Internet, too, has a multilayer structure—so many layers, in fact, that you can easily get lost. Paul Kedrosky, a senior fellow at the Kauffman Foundation, commented that the Internet was supposed to be "The great democratizer of information. It was supposed to empower individual investors, make murky financial markets more transparent, and create a new generation of citizen investors. It was supposed to shrink the world and turn it into a village, where everything happened in the public square and corruption and greed would have no place to hide. As the 1990s mantra went, 'information wants to be free.'"[23]

However, this new "freedom of information," has created a giant jigsaw puzzle comprising a zillion pieces of information that are constantly and frenetically changing. All the relevant information is there, but is impossible to look at in a way that makes sense. The sheer abundance of financial information available online has created a smog of data which makes it difficult—nigh impossible—to grasp the whole picture and comprehend what is going on at a macro level. It is not just information overload that is making the financial system such a precarious place; there is also the issue of impatience. The rise of financial geeks and their derivatives-based models has conditioned banks to expect quick profits, rather than long-term returns. They want instant gratification. This cultural shift has serious consequences—not just for the banking industry—but also for society. According to Andrew Haldane, Executive Director for Financial Stability at the Bank of England, the virtues of patience and delayed gratification have, for centuries, been critical to building sustainable financial, social and economic systems. But now people expect their needs to be met almost immediately. We are becoming a more impatient society. "Impatience is socially, as well as technologically, contagious," Haldane cited the example of the investment trading industry which has recently undergone an exponential leap in terms of speed.[24] "A decade ago, the execution interval for a high-frequency trade (HFT) was seconds. Advances in technology mean today's HFT's operate in milli- or micro-seconds. Tomorrow's may operate in nanoseconds." Some high-frequency traders trade more than a billion shares a day, and they are very short-term bets—by 2013 a typical stock was being held for just eleven seconds.

This hyper speed is compounded by the expectations of investors who are also becoming more impatient. Between 1940 and the mid-1970s, a typical stock in the United States was held for around seven years. This fell to two years by 1987; one year by 2000; and now it is around six months. "Impatience is mounting," lamented Haldane.[25] The cumulative impact of these trends is to further accelerate the speed at which the financial system operates. In the deregulated, frictionless environment of the global economy, money now flows freely between every sector at dizzying speeds on a scale never seen before. No one seems to mind how fast the money flows because it is seen as evidence of "efficiency" and is therefore a sign of progress.

But is it really? Let's look at this from the perspective of the natural world. The ecosystem tells us that there are often serious consequences when a system "speeds up" beyond its natural capacity. Consider, for example, how water moves across a landscape. Professor David W. Orr, the Paul Sears Distinguished Professor of Environmental Studies and Politics at Oberlin College and a James Marsh professor at the University of Vermont, pointed out that if water moves too quickly it does not recharge underground aquifers. The results are floods in wet weather and droughts in summer.[26]

Similarly, Orr said, "Money moving too quickly through an economy does not recharge the local wellsprings of prosperity; whatever else it does for the global economy. The result is an economy polarized between those few who do well in a high-velocity economy [such as the financial geeks] and those left behind. There is an appropriate velocity for water set by geology, soils, vegetation, and ecological relationships in a given landscape. There is an appropriate velocity for money that corresponds to long-term needs of whole communities rooted in particular places and the necessity of preserving ecological capital."

Think for a moment of the staggeringly large volumes of money flowing—at lightning speed—through the big financial institutions in your nearest city, and consider how little of this actually settles into the local business environment. For many people, this is like living next to a giant dam full of water while there is a drought on. The water is not available locally because it is being diverted elsewhere to the big projects on the horizon—always out of reach. The shimmering skyscrapers of

the banks suggest the money is close at hand, but it is not—
for they are simply conduits for a global money pipeline that
is disconnected from much of the rest of the community. The
financial system is a world unto itself.

Meanwhile, attempts by central banks and governments to
prop up failing banks, while maintaining a deregulated system,
driven by hyper-speed trading, will ensure that the really big
busts, like the recent global financial crisis, happen more frequently. There used to be a gap of a few decades between financial crashes, as was the case between the 1929, 1966 and 1987
crashes, and the dot.com crash of 2000. But the recent global
financial crash happened just eight years after the previous
one, and it was a lot bigger. Things are speeding up. As Lawrence Summers, President Barack Obama's principal economic
adviser, has warned, every few years "for the last generation a
financial system that was intended to manage, distribute, and
control risk has, in fact, been the source of risk—with devastating consequences for workers, consumers, and taxpayers."[27]

But can we absorb the lessons? The Nobel Prize–winning
economist Vernon Smith conducted an experiment whereby
some traders were asked to trade an imaginary stock among
themselves. They soon bid up the price of the stock way beyond
its real—fundamental—value until they created a speculative
bubble which finally burst. The traders were then asked to start
over and trade the same stock again, knowing full well what
had happened last time. The result? They did exactly the same
thing again, but this time the velocity at which the bubble was
created was much greater. "We think we can beat the crowd,"
says Smith, "but we are the crowd."[28]

It seems we will not have to wait long for the next global
financial crisis. Meanwhile the computer models—and the geeks

who run them—will continue to accelerate the global financial system to the limits of its capacity, and possibly beyond. It's like an endless game of Domination, but with much higher stakes and no James Bond to save the day.

## *The seeker* (Findus algorithmius)

In 1989, a young surfing enthusiast from Barbados named Alan Emtage was working as a systems administrator at McGill University's School of Computer Science, in Montreal, Canada. "Students were always dropping in asking me to help them find a particular piece of computer software code," he told Melissa Rollock at Nationnews.com. "I developed a set of programs that would go out and look through the repositories of software and build basically an index of the available software. One thing led to another and word got out that I had an index available and people started writing in and asking if we could search the index on their behalf."[29]

But rather than doing this searching themselves, Emtage and his team "wrote software that would allow students to come in and search the index themselves. That was the beginning." When his boss returned from a conference a week later, he asked, "What the hell is going on? I have just been congratulated by a whole lot of people for something I know nothing about." Emtage explained the new file search system he had created. His boss smiled and said, "Well, carry on then."

The search engine was born.

Emtage named his innovation "Archie," which derives from "archive" without the "v". It proved so useful that he and his colleagues soon developed other versions of it for other universities. Over time, explained Emtage, "a lot of people who had used the Internet (and his search engine) at grad schools then went off to work for companies and asked why don't we have access to it?" This prompted corporations to become involved which greatly accelerated search engine technology.

Today we take search engines for granted and consider them to be indispensable. Indeed, it would be impossible to navigate the complexities of the Information Age without them. It would be like looking for a needle in a haystack that is as large as planet Earth.

Like many pioneering geeks, Emtage readily admits there was luck involved in his breakthrough. "I was in the right place at the right time," he told Rollock. "There were other people who had similar ideas and were working on similar projects. I just happened to get there first . . . a lot of the techniques that I, and other people that worked with me on Archie came up with, are basically the same techniques that Google, Yahoo! and all the other search engines use. So, it was my one opportunity to change the world and you don't get many of those."

A decade after Emtage's innovation, two students at Stanford University in California, Sergey Brin and Larry Page, had the brilliant insight that they could create a better search engine by analyzing the relationships between different websites rather than simply counting how many times the search terms appeared on a page. They designed a special algorithm to do this and wrote up their idea in a paper titled "The Anatomy of a Large-Scale Hypertextual Web Search Engine," which has since become the tenth most accessed scientific paper at Stanford University. Next they set up operations in a friend's garage—as Silicon Valley lore seems to dictate—and called their company "Google," derived from a misspelling of the word "googol"—the numeral one followed by one hundred zeroes. One of their early breakthroughs was to figure out how to incorporate advertising into their search engine, because, as Paul Gilster explained in his book *Digital Literacy*, if you were an advertiser, "How could the world beat a path to your door when the path was uncharted, uncatalogued, and could be discovered only serendipitously?"[30] Google solved this problem.

The rest is history. Google now attracts over one billion unique users every month and is by far the dominant search engine in the world.

Why did Brin and Page do so spectacularly well while most other search engines faded ignominiously into the background? They are certainly bright people, even by geek standards. Brin is brilliant at math and Page is highly conceptual. Also, as sons of Jewish academics, they both grew up in environments that fostered their intellectual talents. So their orchid traits would have been well nourished.

Like Alan Emtage, the two men were also in the right place and at the right time—a stimulating campus environment in the United States, with access to lots of technology. It had been a very different story for Brin's parents, who had suffered under the authoritarian anti-Semitic regime of Soviet Russia, where they were prevented from pursuing their career aspirations, despite being amply qualified. It was only when Brin's family migrated to the United States that his father, Michael, was able to become a mathematics professor at the University of Maryland, and his mother, Eugenie, a research scientist at NASA's Goddard Space Flight Center.

"I would have never had the kinds of opportunities I've had here, in the Soviet Union, or even in Russia today," Sergey Brin told Stephanie Strom of the *New York Times*.[31] It is also unlikely—impossible even—that Brin would have had such opportunities if he had been born, say, a few generations earlier, even in America. It would take an entirely new environment to emerge before geeks like Brin and Page could move beyond academic and technical roles, and create a company like Google.

The Anthropocene opened up unlimited possibilities. Indeed, one of the most startling—and somewhat alarming— aspects of the meteoric rise of Brin and Page is the sheer scale of their ambition. They didn't just want to create a more efficient search engine. Their ultimate goal, in their own words, is "To organize the world's information and make it universally accessible and useful." No small task given that, as discussed earlier, a CD stack of the world's data would reach to the moon. The sheer audacity of this vision highlights a recurring theme in this book, which is that super-geeks like Brin and Page don't just want to participate in the Anthropocene, they want to reshape it on their own terms. This ambition is of a different order of magnitude from, say, a CEO of General Motors wanting to create a more fuel-efficient car, or the boss of Airbus committing to build the world's biggest plane. This is about reengineering the Information Age itself.

In doing so, however, tech titans like Google are pushing society to become more numbers driven, automated and machinelike. This can have unintended consequences.

In September 2012, the wife of the former German President, Bettina Wulff, sued the search engine company Google, for linking her to false rumors that she had once worked as a prostitute. Whenever people entered her name in the search box, Google's autocomplete function would offer suggestions like "prostitute," "bordello" or "Playboy." Wulff wrote, in her autobiography, "I have never worked as an escort," and told

*Bild* that the rumors had been very hurtful for her and her family. She didn't want her young son Leander coming across such speculation about her while surfing the Internet.

A month later, halfway across the world, an Australian man, Michael Trkulja, successfully sued Google for linking him to a well-known underworld figure named Tony Mokbel. Trkulja had never been involved in criminal activity but had had the misfortune to be shot in the back in a restaurant during the gangland wars in Melbourne in 2004. Whenever these gangland activities were publicized, Trkulja's shooting would be mentioned. So when people used Google to search for information about the Melbourne underworld, they would see Trkulja's name—even though he had been an innocent bystander.

These two cases highlight one of the darker sides of Internet search engines. Although they are supposed to operate in an objective manner—driven purely by algorithms—they can deliver biased results, and even destroy a person's reputation. This reputational damage can persist for years, perhaps decades, because unlike the old media, today's news is no longer tomorrow's fish and chips wrapping. It persists as a permanent record in cyberspace.

The geeks who create and manage search engines insist that these problems are not their fault. They are just creating algorithms—strings of code—that have no emotional or moral biases. They operate autonomously and objectively. Their argument is strikingly similar to that of the geeks who created the derivatives trading programs that almost destroyed the world's financial system. In the case of Bettina Wulff, a Google spokesperson explained that the associations that come up in relation to her name are "the algorithmic result of several objective factors, including the popularity of search terms."

Technically, this explanation may be true. But the problem is that because search results are largely based on which sites are most popular, this creates a snowballing effect. The more people who visit a particular site, the higher it is ranked, which in turn, ensures more people visit that site. It is a self-reinforcing circle. This effect is compounded by the fact that people are naturally drawn toward sensational tabloid-style news items that involve notoriety. Over time, this ensures that the search engine rankings are often biased toward the negative. This dual nature of search engines means they can both build and destroy reputations.

Search engines can also exert a subtle—yet profound—influence on the way we see the world by narrowing our focus. This is because they operate like a laser, cutting through the information clutter, to find exactly what we are looking for. A consequence of this laser-like focus is that it can encourage "confirmation bias," whereby we look for information that confirms our existing mindsets and prejudices. When we go online, a search engine prompts us to act as our own editor and gatekeeper for the news and information, and to screen out opposing viewpoints. Think for a moment: how often do we search for viewpoints that contradict our own?

Nicholas Negroponte of MIT called this self-censored news "The Daily Me" and is convinced it represents another step toward a world in which people increasingly isolate themselves in a bubble of self-sustaining beliefs, or emerge to immerse themselves in like-minded communities (information cocoons). Although we may think we like the idea of a debating chamber, in reality we prefer an echo chamber.[32]

In one classic US study, Republicans and Democrats were offered various research reports from a neutral source. Both groups were most eager to receive coherent arguments that corroborated their pre-existing mindsets. Bill Bishop, the author of *The Big Sort: Why the Clustering of Like-Minded America Is Tearing Us Apart*, has said that as the United States grows more politically segregated, "the benefit that ought to come with having a variety of opinions is lost to the righteousness that is the special entitlement of homogeneous groups."[33] A twelve-nation study found that Americans, particularly highly educated ones, are the least likely to discuss politics with people of different views.

Which brings us back to the geeks—that most highly educated of all groups. You would think that, given their natural curiosity and insatiable appetite for data, geeks of all people would be most eager to seek out a wide range of views. But as we saw earlier in "What makes a geek?" they tend to drill down deeply into their chosen subject (or obsession). Their concentrative power is what makes them so good at what they do. They don't dilute their efforts.

This narrowness of perspective may help explain why Google has been reluctant to address the wider problems arising from its search engine technologies, such as the reputational damage to individuals like Mrs. Wulff and Mr. Tjkulja. They are so enamored with their algorithms they cannot see the collateral damage being inflicted. They don't mean to cause harm, of course. In fact, when Google published its initial public offer document in 2004, it featured a manifesto titled "Don't be evil," and included the statement, "We believe strongly that in the long term, we will be better served—as shareholders and in all other ways—by a company that does good things for the world even if we forgo some short term gains."[34]

The trouble is that over time the more powerful an organization—or system—becomes, the higher the probability it will cause unintended damage. This is because the needs of the system begin to transcend the needs of the people it serves. What starts off as a noble idea eventually devolves into a self-serving one. I am not suggesting that Google has reached that point yet, nor is anywhere close to it. But none of us wants to live in the kind of society in which an algorithm can determine the value of a person's reputation. It would be like living in the days of the Soviet Union when a faceless bureaucrat could determine the fate of a brilliant young academic.

## The cyber geek (Hackorum neophilum)

Jonathan Millican is not the sort of person you would expect to find on the front lines of a global war. At nineteen years of age, with his tousled brown hair, he looks like the typical university student that he is. Yet, in a series of grueling challenges he demonstrated a ruthless ability to defend his side against an unseen enemy and hunt down rogue invaders.

Millican is the winner of the 2012 Cyber Security Challenge, which is a UK-government initiative designed to attract more talent to the fight against cyber threats. He beat off about 4,000 other bright young contestants who spent six months hunting software viruses, launching firewalls and repelling fake attacks in a series of simulations. Although he does not consider himself a "geek," Millican's fellow students at Cambridge University—where he studied computer science at Jesus College—might beg to differ.

In presenting the winner's award at a ceremony in Bristol, Adam Thompson, the chief judge, said Millican, "showed great leadership, strong technical abilities and demonstrated that he understood the impact what he was doing would have on a business. . . . He demonstrated knowledge years beyond his time."[35] In response, the young computer scientist made it clear that he appreciated what was at stake. "We're going into an age of cyber warfare," he said, and "Given all the critical systems we have in this country that are connected to the Internet it's very important that there are experts out there that can keep people safe."[36]

There is no doubt that our world has become a lot more dependent on computers in recent years. Nearly all our modern infrastructure, including energy supplies, air-traffic control, transport grids, water supplies, and communication, is controlled by interconnected computer networks, which are vulnerable to cyber-attacks. Each year in the United States alone, there are about 50,000 incidents of malicious cyber activity, and every day computer systems are searched by millions of automated scans from foreign sources trying to find unprotected communications ports. Hackers even gained temporary control of the NASA's Jet Propulsion Laboratory in Pasadena, which manages numerous spacecraft, including the International Space Station. One Washington think tank put the cost of cyber-attacks in the United States last year at roughly US$100 billion. Most other advanced countries are similarly affected. A survey commissioned by the UK Cabinet Office estimated that the British economy lost £27 billion to cyber theft in 2010.[37]

In a speech at the Intrepid Sea, Air and Space Museum in New York in 2012, the US Defense Secretary, Leon Panetta, warned that the United States was facing the possibility of a "cyber Pearl Harbor." "An aggressor nation or extremist group could gain control of critical switches and derail passenger trains, or trains loaded with lethal chemicals," he said. "They could contaminate the water supply in major cities, or shut down the power grid across large parts of the country."[38]

Hackers have been able to penetrate networks for months, or even years at a time, without being detected, according to Shawn Henry, Executive Assistant Director at the FBI. He told *The Guardian* newspaper that these hackers would "have essentially had free rein over those networks . . . with the ability not only to review that data, but potentially to change that data . . . (and) to disrupt that network entirely."[39] Henry compared the situation to a game of chess, in which, "every step that the defense makes, the offense changes its tactics." It is not just a defensive game, of course; sometimes, there is a need to go on the offensive by creating programs such as the now-famous Stuxnet worm which attacked Iran's nuclear systems.

The work of the FBI and other Western agencies would be a lot easier if the best cyber warriors all worked for them. In fact, most of them did so up until a few decades ago. Only Western nations could afford to educate and develop potential cyber sleuths like Jonathan Millican. Not even the Soviet Union in its heyday could match the West in this respect. But it is now a much more level playing field. Rising powerhouses like China and India can afford—and are motivated—to create their own cadres of computer hackers who are capable of uncovering commercial secrets or waging war. Criminal groups have also entered the fray. In June 2012, the passwords of over six million LinkedIn accounts were posted on a Russian hacker forum. A month later, in July 2012, Yahoo announced that 450,000 of its user names and passwords had been compromised.

The biggest security breach of all occurred in early 2014, when it was revealed that about 17 percent (over half a million) of the Internet's so-called secure webservers were vulnerable to attack. This security bug, known as "Heartbleed," was described by *Forbes* cyber security columnist Joseph Steinberg as "the worst vulnerability found (at least in terms of its potential impact) since commercial traffic began to flow on the Internet."[40]

It is increasingly clear that, despite advances in cryptography and firewall protections, it is impossible to make a computer network 100 percent secure from an attack by a determined hacker.

So what kind of person is capable of wreaking so much devastation in cyberspace?

## Portrait of a hacker

The simple answer is that most of them are geeks—that is, acutely intelligent young people endowed with a laser-like focus. It doesn't really matter whether they are American, Chinese, British or whatever—what distinguishes them is their ability to make sense of—and cut through—complicated digital networks that would baffle most of us. Like other geeks we have met in this book, such as in the financial and tech communities, these people prefer dealing with synthetic versions of reality. Many of them are sensitive "orchid types" who find themselves "switched on" by the competitive hothouse environment of the hacker community.

One of the most comprehensive attempts to define the hacker personality can be found in *A Portrait of J. Random Hacker*, which was created by The Cyberpunk Project, and is extracted here as follows:

"The most obvious common 'personality' characteristics of hackers are high intelligence, consuming curiosity, and facility with intellectual abstractions. Most hackers are 'neophiles,' stimulated by and appreciative of novelty (especially intellectual novelty). Hackers can often discourse knowledgeably and even interestingly on any number of obscure subjects—if you can get them to talk at all, as opposed to, say, going back to their hacking. Hackers tend to be careful and orderly in their intellectual lives and chaotic elsewhere. Their code will be beautiful, even if their desks are buried in 3 feet of crap. Their reading habits are omnivorous, but usually include a lot of science and science fiction. Hackers often have a reading range that astonishes liberal arts people but tend not to talk about it as much. Many hackers spend as much of their spare time reading as the average American burns up watching TV."

Nearly all hackers, past their teens, are either college-degreed or self-educated to an equivalent level. The self-taught hacker is often considered (at least by other hackers) to be better-motivated, and may be more respected, than his school-shaped counterpart. Politically, hackers are vaguely left of center, except for the strong libertarian contingent which rejects conventional left-right politics entirely. The only safe generalization is that hackers tend to be rather antiauthoritarian. Hackers have relatively little ability to identify emotionally with other people. This may be because hackers generally aren't much like "other people." Hackers tend to be especially poor at confrontation and negotiation.

As a result of the above traits, many hackers have difficulty maintaining stable relationships. At worst, they can produce the classic computer geek: withdrawn, relationally incompetent, sexually frustrated, and desperately unhappy when not submerged in his or her craft. Fortunately, this extreme is far less common than mainstream folklore paints it—but almost all hackers will recognize something of themselves in the unflattering paragraphs above.[41]

The most famous—or infamous—hacker in the world is an Australian, Julian Assange, the founder of Wikileaks, which is a site where people can upload secret information and make it public. In 2010 Wikileaks released hundreds of thousands of classified US diplomatic cables about the Iraq War and various intergovernmental relationships. These documents had a seismic and polarizing effect on the global public consciousness, and even helped to inspire political movements, particularly in the Middle East. Assange himself came under intense political and media pressure, and even faced death threats for his actions, which some Americans considered treasonous. This pressure was compounded by some messy complications in Assange's personal life that eventually led to an extradition standoff between the governments of the UK, Sweden and Ecuador. The man who had wanted to change the world now became the center of attention for all the wrong reasons. Whether history will judge Assange to be a constructive or destructive force, it is too early to tell, but his impact has certainly been profound.

Long before he rose to global prominence, Assange was featured in the 1997 book by Dr. Suelette Dreyfus called *Underground: Tales of Hacking, Madness and Obsession on the Electronic Frontier*. Dreyfus interviewed dozens of hackers around the world, including Assange—whose codename then was Mendax—about their lives and motivations.[42] These interviews reveal a polyglot cast of brilliant eccentrics, who were keen to make an impact on the world, while keeping their own identities secret. For his epigraph in the online edition of Underground, Assange quoted Oscar Wilde: "Man is least himself when he talks in his own person. Give him a mask, and he will tell you the truth." In effect, this is what Wiki-leaks does—provides a mask for people to reveal their "truth" by enabling them to anonymously deposit documents in an encrypted digital vault, which are then disseminated widely. But as Assange discovered, the disclosure of some truths can have serious consequences.

Beyond the political repercussions of the Wikileaks revelations, the Julian Assange case highlights a fundamental difference in the way geeks and non-geeks view the world, and specifically how "facts" are treated in the information age. Geeks tend to see "facts" as valuable or interesting in themselves. That is, they believe that facts don't necessarily require context to give them meaning or relevance. They speak for themselves, and have their own intrinsic meaning. Therefore, geeks believe that facts should be disseminated and made available to as many people as possible—in their raw state. It is up to the recipient to make what they will of them. This behavior is consistent with how some people on the autism spectrum communicate. They tend to blurt out facts in a way that can come across as emotionally blind, and with little regard for the consequences.

On the other hand, non-geeks (which is most of us) prefer facts to be contextualized within a narrative structure or story. They like them to be interpreted—or pre-digested—so they can better comprehend their meaning. They want the filter. In our society the main filters are the established media companies, whose job it is to sift through the myriad facts available and make sense of them. They employ professional journalists who are trained to interpret the news, in accordance with an agreed set of criteria that includes the public interest, morality, legality, taste, and so on. So the raw information is never simply dumped into the public domain; it goes through a refining process.

The shock of the Wikileaks revelations wasn't so much to do with the Machiavellian nature of US policy—which was not surprising given the nature of international relations—but that such information could bypass the traditional "refining" process. We have become so accustomed to having our information prepared and served on a platter that we weren't used to receiving it in its unadulterated form. Certainly, a few major media organizations became involved in the Wikileaks dissemination process, but more as an adjunct than as filters. Most of the diplomatic cables could be read directly online in their original form.

The difficulty in judging the Wikileaks saga has much to do with the issue of intent. Because, unlike most crimes, the protagonists believed that they were acting in the public interest. As geeks, they thought it was their duty to get the "facts" out there. Similarly, other hacker-based organizations such as Anonymous—which has shut down major websites and orchestrated mass protests—say they are motivated by a desire to increase the level of public transparency and to protect civil liberties. Gabriella Coleman, the Wolfe Chair in Scientific and Technological Literacy at Canada's McGill University, and a trained anthropologist, believes that hackers, coders and geeks are behind a vibrant political culture. "A decade-plus of anthropological fieldwork among hackers and like-minded geeks has led me to the firm conviction that these people are building one of the most vibrant civil liberties movements we've ever seen." she wrote in the *MIT Technology Review*, "It is a culture committed to freeing information, insisting on privacy, and fighting censorship, which in turn propels wide-ranging political activity. In the last year alone, hackers have been behind some of the most powerful political currents out there."[43]

Most hackers, however, are not driven by the pursuit of a noble purpose, but by intellectual curiosity and stimulation. They are neophiliacs—people with a strong desire for novelty. As one hacker who goes by the name of Weld Pond told the US Public Broadcasting Service, "It's actually a sort of thrill finding the problems, the thrill of exposing the weaknesses and saying, 'Well, geeze, look, they had all these smart people design this system, and I spent a few hours and I looked at it, and look, I found this huge problem.' That's kind of exciting."

This intellectual adventurism, however, can cause some hackers to end up in the courts. When Gary McKinnon, a British man with Asperger's syndrome, was caught breaking into sensitive US military and NASA installations from his flat in 2002, he claimed he was looking for evidence of UFOs. It didn't help his case that he had posted a notice on the military's website, saying, "Your security is crap."

The US authorities cracked down hard on McKinnon, stating that he deleted critical files from operating systems, which shut down the US Army's Military District of Washington computer network for twenty-four hours. They also claimed that he deleted weapons logs at the Earle Naval Weapons Station shortly after 9/11, which had the effect of paralyzing munitions supply deliveries for the US Navy's Atlantic fleet. To deter other would-be hackers, the US government tried to extradite McKinnon to face US justice. But the attempt eventually failed following the intervention of the UK Home Secretary, Theresa May, in October 2012, after a decade of legal and political wrangling, and a public outcry at McKinnon's treatment. May decided that McKinnon was too "ill" to be extradited to the United States and blocked the extradition on the ground of human rights.

It turns out that McKinnon is just one in a long list of people diagnosed with Asperger's who have been caught hacking. According to Misha Glenny, the author of *Dark Market: How Hackers Became the New Mafia*, such people can be drawn to nefarious online activities because of their prodigious computer knowledge and lack of social skills. "The Internet is what made them," he wrote, "it gave them a place to be criminals." Glenny argues that they were predisposed to getting tangled up in the dark side of the web.

Another hacker who became entangled is Ryan Cleary, who was arrested by Britain's Serious Organised Crime Agency in March 2011, on suspicion of being one of the masterminds behind the LulzSec group, which hacked into the computer systems of the CIA, US Senate, US Air Force, Sony Corporation, Nintendo and the UK National Health Service. The name LulzSec is a combination of lulz or lols, meaning "laugh out loud," and sec for "security." Detectives raided Cleary's Essex home and nabbed him in his bedroom, to which he had virtually confined himself over the past four years. His mother said later that her then nineteen-year-old son was an agoraphobic recluse, and only left the bedroom to use the bathroom. She would leave his food outside the door. Cleary's twenty-two-year-old half-brother told the *Sun* newspaper, "Ryan is obsessed with computers. He's a bit of a geek. That's all he does—he's a recluse. He locks himself in his room every day, closes the curtains and spends hours at a time online."[44]

People like Ryan Cleary and Gary McKinnon, personify the asymmetric nature of digital technology, whereby an ingenious individual can wield enormous power through a few strokes of a keyboard. Some observers believe that it is a shame that such prodigious talents are not harnessed for more constructive purposes, and suggest that society should not be too harsh on hackers.

John Arquilla, a professor of defense analysis at the US Naval Postgraduate School in Monterey, California, told the *Guardian* newspaper, that the US government should think about hiring rather than prosecuting hackers like McKinnon. "There are other places in the world where these communities are embraced by official authority, and these are places that are becoming great cyber powers. The analogy is as if after World War Two, the Russians were using these rocket scientists while we put the ones we got on trial and incarcerated them."[45] One of the problems with hiring hackers, however, is that their background in illicit activities makes it hard for them to be given security clearance. "We have to create a new kind of institutional culture that allows us to reach out to these diverse kinds of actors," Arquilla said.

This new culture may be a long time coming, however, not just because of legal barriers, but because of an entrenched social bias against anyone who is considered to be non-neurotypical — as geeks often are. Even security agencies, for all their cloak and dagger activities and operational flexibility, insist that their employees fit into the system. Renegades are not welcome.

Meanwhile, the cyber arms race is gathering pace. "It's not at a point where I would call it cyber war yet, but it's close," said Larry Clinton, President of the Internet Security Alliance, which represents many businesses, including defense contractors. "I think we are certainly seeing an arms race with respect to cyber. We did well to get through the nuclear age. We did well with chemical weapons. If we can do as well with cyber, that would be great, but we don't really have a theory; I am not sure what the theory is. We don't have a model set up for how we are going to deal with this."[46]

The essential problem is that the very qualities that make the Internet such a valuable tool—its accessibility and interconnectedness—also make it especially vulnerable to a cyber-attack. And unlike a traditional military attack, which requires a build-up of weapons and logistics, a cyber-attack can happen instantaneously—without a hint of warning. Yet the damage inflicted can be every bit as consequential, because its effects radiate throughout the network, and are not confined to physical infrastructure like a bridge, or government building. Vital networks could crumble under the onslaught, and massive blackouts would bring major population centers to a standstill. Modern cities would be forced back into the Dark Ages, stripped of even the most rudimentary utilities. Our economic "assets," which consist mostly of records in bank computer systems, could be permanently erased. Incalculable amounts of vital information would be obliterated. The damage would not just be physical but psychological, and deeply traumatic to societies that have grown so utterly dependent on modern technology.

To put it bluntly, cyber warfare threatens the very foundations of the Anthropocene. Paradoxically, our collective security now lies in the hands of the very geeks who created these vulnerable networks in the first place.

## *The Amazonian* (Omnia discountia)

In August 2011, a group of Brazilian scientists announced a remarkable discovery. Deep beneath the mighty Amazon River is another river, which flows underground along the same route for over 6,000 kilometers before emptying into the Atlantic Ocean.[47] The scientists presented their findings to the Twelfth

International Congress of the Brazilian Geophysical Society in Rio de Janeiro, and named the river "Hamza" in honor of its co-discoverer, Dr. Valiya Mannathal Hamza, a professor of geophysics at the Brazilian National Observatory. They explained that the Hamza operates as a giant aquifer, whose water helps underpin the fragile ecosystem of the Amazon Basin.

Like many natural phenomena, the existence of the Hamza River has something to tell us about our man-made world. It reminds us that there is often more going on than meets the eye.

Which brings us to the "other Amazon"—the one that gushes out an endless stream of books, music, films, software, Kindles and other products. Since it was launched in 1995, Amazon.com has grown into a colossus of the online world. By 2012, it generated an annual revenue of nearly US$50 billion, which was more than the entire global publishing industry combined for that year.[48] Like the river it was named after, Amazon has developed a seemingly unstoppable momentum and has become a household name.

Despite its huge footprint, most of us have a one-dimensional view of Amazon because we only experience it through a computer screen or smartphone. What we don't see is the vast technological infrastructure, which—like the Hamza River—supports the Amazon online ecosystem. This infrastructure comprises data centers built to hyper-scale, with extraordinary processing power, inventory management systems, communications networks and management processes.

Increasingly, much of this infrastructure operates "in the cloud," in that it is accessible from anywhere through the Internet. In fact, Amazon's cloud computing services not only support Amazon, but much of the corporate world too. So, for example, instead of building its own data storage facilities, a company can use Amazon's cloud service. It is not just businesses that use these services. In 2012, President Obama's campaign team used Amazon's service to help coordinate hundreds of thousands of staffers and volunteers, various donation programs, and the Get Out the Vote campaign. "It [Amazon's Cloud] made a big difference to the success of the campaign," explained Michael Slaby, chief integration and innovation officer for the Obama team.[49]

The point is Amazon is no longer just an online retailer. It has evolved into one of the pillars of the Anthropocene, whose capabilities can influence even a US election.

## The backroom guy

Jeff Bezos, the founder of Amazon, doesn't enjoy the same high public profile as the Four Horsemen of the Digital Age—Mark Zuckerberg, Sergey Brin, Bill Gates and the late Steve Jobs. One reason for this is because he is seen by industry insiders as a "backroom guy," that is, someone focused on building infrastructure. For example, after surviving the dotcom crash in 2001, he realized that his data centers were using less than 10 percent of their capacity at any one time, excluding those periods when there was a spike of activity. So his team set about reengineering this excess capacity to make it available to other customers online, and adding new functions. This initiative heralded the beginning of Amazon's "cloud computing" business, which now accounts for the fastest growing part of its revenues.

Like his more famous digital brethren, Bezos also has his quirks—such as a boisterously demonic laugh, and a penchant for regulation Oxford blue shirts and jeans. He is a stickler for detail too. The online magazine Portfolio.com described him as, "at once a happy-go-lucky mogul and a notorious micromanager . . . an executive who wants to know about everything from contract minutiae to how he is quoted in all Amazon press releases."[50]

Another characteristic that Bezos shares with the super-geeks we have met in this book is that he harbors a gargantuan ambition. This is to create a world in which nearly every new book, and much else as well, is sold through his online distribution system. He also wants much of the business world to depend on his technological infrastructure—his underground "Hamza." Bezos first became acquainted with the power of technology as a toddler, when he tried to dismantle his crib with a screwdriver. Some years later he rigged up an electric alarm to keep his younger siblings out of his room. As a student he proved to be brilliant; graduating from Princeton University summa cum laude, Phi Beta Kappa, with a Bachelor of Science in electrical engineering and computer science. Next came a few years working as an investment banker on Wall Street. Then he launched Amazon, in that traditional "birthing center" of all big tech companies, his garage.[51]

The first book Bezos sold on his fledgling Amazon site in 1995 was Douglas Hofstadter's *Fluid Concepts and Creative Analogies: Computer Models of the Fundamental Mechanisms of Thought*. It was a good read for geeks, no doubt, but an inauspicious start for what would become the world's largest bookseller. Indeed, since its launch, Amazon has made more books available to more people than any organization in history—even more than Gideon's has through its free Bible service. It is fair to say that Amazon is responsible for the biggest change to publishing since the invention of the Gutenberg press over four hundred years ago.

The meteoric rise of Amazon, however, has not been welcomed by everyone. Bezos has seriously disrupted some long-established industries, particularly publishing and retailing, and dismantled their business models in much the same manner as he once attacked his childhood crib. Not that this concerns him, as he told *Wired*'s Steven Levy in November 2011, "As a company, one of our greatest cultural strengths is accepting the fact that if you're going to invent, you're going to disrupt. A lot of entrenched interests are not going to like it. Some of them will be genuinely concerned about the new way, and some of them will have a vested self-interest in preserving the old way. But in both cases, they're going to create a lot of noise."[52]

This noise became louder in 2010, when Amazon's share of the book market topped 70 percent in many countries, and some major book retailers, such as Borders and Barnes and Noble, were forced to close—blaming Amazon's aggressive pricing for their demise. Traditional booksellers, who based their businesses on nurturing relationships with customers, could not compete with the machinelike efficiency of Amazon, which offered hardly any nurturing at all. "Our version of a perfect customer experience," Bezos told Steven Levy in the *Wired* interview, "is one in which our customer doesn't want to talk to us. Every time a customer contacts us, we see it as a defect. I've been saying for many, many years, people should talk to their friends, not their merchants. And so we use all of our customer service information to find the root cause of any customer contact. What went wrong? Why did that person have to call? Why aren't they spending that time talking to their family instead of talking to us? How do we fix it?"

By eliminating the middleman—the retailer—Bezos has been able to drastically lower the price of books (often below $10) and offer an almost unlimited choice. He explained to Levy that he would prefer to have smaller margins and a bigger customer base than big margins and a smaller customer base.

From a purely business perspective, Amazon's numbers-driven model makes perfect sense. But as Bezos learned a long time ago, it is not always about the numbers. In a speech to graduates at his alma mater, Princeton University, in May 2010, he recounted a childhood incident. While sitting in a car with his grandmother, who was a heavy smoker, he calculated by how many years her addiction would reduce her life expectancy. When he announced the result from the back seat, he expected praise for his clever math ability. But instead his grandmother burst into tears.[53]

Many booksellers have burst into tears in recent years. They wring their hands in despair at customers who browse their shops, find a book, and go home to buy it more cheaply on Amazon. Richard Russo, writing in the *New York Times*, cited the example of Lacy Simons, who ran a small bookshop called Hello Hello near the coast of Maine in the United States. In her blog, Simons wrote about the kind of relationship she aspires to have with the people who visit her store:

"If you let me, I'll get to know you through your reading life and strive to find books that resonate with you. Amazon asks you to take advantage of my knowledge & my education (which I'm still paying for) and treat the space I rent, the heat & light I pay for, the insurance policies I need to be here, the sales tax I gather for the state, the gathering place I offer, the books and book culture I believe in so much that I've wagered every-thing on it as if it were showroom for goods you can just get more cheaply through them."[54]

As if to rub salt in the wound, in 2011, Amazon launched a new promotion inviting customers to visit their local bricks-and-mortar bookstore and use its new price-check app to scan product barcodes and compare them with Amazon's prices. Many people, including some famous authors, thought Amazon had gone too far. "Scorched earth capitalism," is how Dennis Lehane described it to Richard Russo, for the same *New York Times* article. Another big-name author, Steven King, said that while he loved his Kindle, he thought the practice of using the price-check app was "invasive and unfair." Scott Turow, president of the US Author's Guild and a practicing lawyer, wrote, "The law has long been clear that stores do not invite the public in for all purposes. A retailer is not expected to serve as a warming station for the homeless or a site for band practice. So it's worth wondering whether it's lawful for Amazon to encourage people to enter a store for the purpose of gathering pricing information for Amazon and buying from the Internet giant, rather than the retailer. Lawful or not, it's an example of Amazon's bare-knuckles approach."

## The limits of free choice

We can't blame consumers—or ourselves—for being attracted to Amazon's low-price model. It is all about freedom of choice. But we should not kid ourselves that there is no downside to this way of doing business, because true freedom also means choice of supplier, not just product. The fact is that our world is increasingly dominated by a few major online "gateways" such as Amazon, Apple, Sony and Microsoft, which control access to the largest range of books, music, or whatever. Not only that, when we purchase, say, an e-book or music track through these gateways, we don't have full ownership of it because the copyright is digitally protected, so we can't share it with anyone else, or resell it. We are forced to use our products within the digital walled gardens of the companies who sell them. So, for example, we can only read an Amazon-purchased book on a Kindle, or an iBook on an iPad. If we move to another platform, say from a Kindle to a Sony reader, we lose access to all the books we bought on the original platform. Harking back to the old industrial days, it is as if Henry Ford didn't just sell cars, but also told us which roads to drive them on. If we moved onto a different highway system our car would be confiscated. Similarly, these digital gatekeepers are saying, "You can have any product you like as long as it is one of ours, and you can only use it on our systems."

Digital rights management, which protects files being copied, is necessary, of course, to prevent people from downloading creative content, such as music, videos and books, without paying for it. It is an anti-piracy measure to protect intellectual copyright. This is why most big publishers supported the introduction in the United States of the Digital Millennium Copyright Act in 1998. The trouble is that once the Act was passed, publishers effectively outsourced the digital rights management to the big online gateways, such as Amazon, whose in-house technicians each developed their own digital locks. Critics of DRM, such as the Electronic Frontier Foundation, argue that it should not be called digital "rights" management, but digital "control" management, and that DRM is essentially an anti-competitive practice, because it stifles the freedom of consumers by locking them into a small number of suppliers.

By putting technicians — that is, geeks — in charge of digital copyright, our options for accessing creative works has been reduced. This highlights one of the great paradoxes of the Information Age. Never before have we, as a society, been exposed to so much information, yet we have so few ways to actually access and manage it. We have become increasingly dependent on just a few key brands, usually Google, Apple, Facebook and Amazon, to navigate, interact and transact with the Anthropocene. Not only that, but the digital infrastructure that supports these brands is becoming increasingly concentrated in fewer hands. This trend will accelerate as more of our vital information is held "in the cloud" through Internet-based computing services. This is not a deliberate Orwellian plan, of course, but rather the natural evolution of digital technology, and our society's obsession with convenience and efficiency. It is so much easier to go with the flow.

Indeed, we all now live downstream of a mighty technological river — which flows relentlessly forward like the Amazon — and sucks us into its current. At the head of this river are a handful of geeks, who have the power to set the river's course and direction. And they are doing just that.

## The warrior geek (Terminus remoticus)

Colonel D. Scott Brenton guides his aircraft over the rugged terrain of Afghanistan in order to get another look at the compound below, where militants are suspected of hiding. He and his team have been watching this place for months. It is not just militants who operate in this area, however; it is also civilians. "I see mothers with children. I see fathers with children,

I see fathers with mothers, I see kids playing soccer." Brenton patiently waits for a window of opportunity to strike his target, but only when the women and children are not around. The risks must be weighed up quickly. He is used to this sort of pressure, though, because as a trained F-16 fighter pilot he has conducted many missions during his fifteen-year career.[55]

But it is different now, because Brenton is not flying in an F-16, and is nowhere near Afghanistan, nor, for that matter, any other war zone. He is 11,000 kilometers away sitting before a computer console in a leafy suburb of Syracuse, New York, explaining his day job to Elisabeth Bumiller, the Pentagon correspondent for the *New York Times*.

Brenton is the pilot of an MQ-9 Reaper drone—a lethal unmanned aircraft designed by General Atomics that is capable of hovering for extended periods over a target at an altitude of 50,000 feet (15,000 meters), before unleashing a Hellfire missile or laser-guided bomb. He works with his partner, a sensor operator, who controls the cameras. Many operators are ex-fighter pilots, like Brenton, but not all of them, explained Bumiller in her article. Some flew C-130 transport planes in Iraq or Afghanistan. Some have seen real combat, others not. In fact, many drone pilots have no conventional flight training at all, and have developed their skills entirely on the drone joysticks which resemble video game controls.

What they all have in common, though, is that they represent a new generation of military personnel—the warrior geek. People who use advanced technical systems to wage war, without exposing themselves physically. They operate in a twilight world between the virtual and the real, where, even though they are dealing with simulations and video feeds, their actions have lethal consequences.

Brenton revealed to Bumiller that there is a peculiar disconnection between real action and fighting a war with a joystick from his comfortable seat in an American suburb. He recalled that when he was deployed in Iraq, "you land and there's no more weapons on your F-16, [so] people have an idea of what you were just involved with." Now he said, "he steps out of a dark room of video screens, his adrenaline still surging after squeezing the trigger, and commutes home past fast-food restaurants and convenience stores to help with homework—but always alone with what he has done."

Technology has always been a major factor in war, of course, ever since our hominid ancestors started bashing each other with rocks and wooden clubs. Weapons development has progressed in leaps and bounds ever since. During all this time, however, warriors have retained a physical connection to the battleground. Whether they were soldiers, sailors or airmen, they were expected to put their lives "in harm's way." This element of danger meant that there were relatively few people in a society who were capable of becoming warriors. It was a job that required physical courage. The sheer physicality of war also helped to ensure that people were acutely—and often painfully—aware of the consequences of their actions.

Remote warfare changes this equation. According to Dr. Peter Singer, who previously headed Barack Obama's defense policy team and is the author of *Wired For War*, "We are living through the end of humankind's 5000-year-old monopoly on the fighting of war. . . . The robots of today are the first technologies to change the "who" of war, not just the "how" of war . . ." In other words, these new remote controlled weapons—which are becoming more intelligent, autonomous and technically demanding—are changing the type of warrior needed to operate them. In an age of joysticks, physical courage is becoming less important than other characteristics, such as visual and cognitive dexterity and the ability to absorb large quantities of fast flowing data. These are, of course, typical geek traits.

Indeed, over time, there will be fewer, and eventually zero, drone pilots who have experienced real-life combat. Their entire service will have been spent in an air-conditioned room, a coffee cup by their side, and located just a few miles from their suburban home. They will never have tasted the adrenaline-fuelled rush of war—with its frenetic cacophony of sights, sounds and smells. Nor will they have felt the gnawing fear, or the revulsion at the casualty-strewn aftermath of a fire-fight.

This evolution from warrior to geek has prompted some old-school military personnel such as Sir Ian Burridge, a former British air chief marshal in Iraq, to call the drone war "a virtueless war," requiring neither courage nor heroism. Indeed, remote warfare can elicit in some people an exaggerated sense of personal power, or even omnipotence. Recall from the Introduction the example of the young video game player who would be transformed into Genghis Khan while playing Space Invaders, only to revert to his more fragile self once he stepped outside into the real world. In an actual battlefield, with bullets whizzing by, no soldier has delusions of being Genghis Khan.

When the US Defense Secretary Leon E. Panetta announced in February 2013 the creation of the new Distinguished Warfare medal—known as the Drone medal—in recognition of the extraordinary achievements that directly impact on combat operations, but that do not involve acts of valor or physical risk that combat entails (i.e., piloting a Predator or Reaper drone from a remote trailer), it was greeted with a barrage of criticism. The Pentagon's decision to rank the Drone medal above the Purple Heart and Bronze Star was particularly galling to many people, and resulted in an online petition at Whitehouse. gov, which stated:

"Bronze Stars are commonly awarded with a Valor device in recognition of a soldier's service in the heat of combat while on the ground in the theater of operation. Under no circumstance should a medal that is designed to honor a pilot, that is controlling a drone via remote control, thousands of miles away from the theater of operation, rank above a medal that involves a soldier being in the line of fire on the ground."

Eventually, the firestorm of controversy over the Drone medal forced the Pentagon to backtrack and replace it with a different category of award, thus ensuring that only those who risked all in combat would be eligible for the higher ranking medals.

This is not to suggest that the people operating the drones do not face consequences or feel stress. They certainly do. Although they are sitting a long way from the frontline, the high quality of their video feeds ensures that they are very much "there" in the action. They are confronted with the results of their work. Also, if they have been monitoring a target for a long period of time, the operators may have become intimate with the daily routines of their targets—including all the quotidian activities that can remind them of their own lives—the family duties, gardening, mucking around with the kids. So when it comes time to hit the "fire" button, it can be difficult for them to do this. Many drone pilots learn to compartmentalize their job, so as to minimize their emotional response, but not everyone can do this. Will, an air force officer who was a drone pilot at Creech air force base in Nevada, explained to Bumiller, "There was good reason for killing the people that I did, and I go through it in my head over and over and over.

But you never forget about it. It never just fades away, I don't think—not for me."

It is perhaps not surprising that an increasing number of drone operators have become afflicted by Post Traumatic Stress Syndrome (PTSD). One theory for this is that their bodies are not given an opportunity to "somatically process" the experience of battle in a physical way. According to the trauma expert Dr. Peter Levine, artificial interfaces (computer screens) change the way our nervous system engages with reality, and therefore subvert our ability to deal with emotions. So, for example, a drone operator who unleashes hell on a group of enemy combatants will experience the consequences of his actions at a *cognitive* level, but has no way of processing it through the body's "felt sense"—which is designed to mediate our emotional responses. This problem can be compounded by conflicting feelings a drone operator may have about their mission, resulting in what is known as a "moral injury"—a term coined by the clinical psychiatrist Jonathan Shay.

## The only game in town

Notwithstanding the moral and ethical issues involved in remote warfare, the use of military drones will continue to increase exponentially. In 2001, the US Air Force operated just seven drones—now there are over 10,000 in service, including Predators, Reapers and Global Hawks. They are controlled by thirteen bases spread across the United States, employing 1,300 drone pilots, and there are more in the pipeline. The CIA also operates its own classified program in places including Pakistan, Somalia and Yemen, which targets "high value" terrorists. Over the past five years

President Obama has authorized hundreds of drone missions in various war zones, in pursuit of enemy insurgents. And it's not just the United States involved. Other countries too are developing drone programs, including Russia, China, the UK, Israel, and dozens more. A number of defense industry analysts suggest that the current crop of fighter planes under development—such as Lockheed's fifth generation F-35—may be the last manned combat planes ever built. Future fighter jets will all be drones.[56]

There are good reasons why drones have become the new weapon of choice for governments of all persuasions. Firstly, they are lethally precise, and can spot a milk carton from sixty thousand feet. This accuracy means less collateral damage. Secondly, they reduce the need to put soldiers in harm's way, which is important for war-weary nations. No leader wants to have to write condolence letters to the loved ones of deceased soldiers. Finally, the drones—being secret programs—operate out of public view. Even when the media do cover drone activity, they often report it in a manner more appropriate to the information technology section of a newspaper. That is, replete with statistics but minus the human element.

Drones can also provide valuable cover for soldiers on the ground by looking out for enemy forces. Because they can stay aloft all night and have infrared sensors, they can even act as night guards for the exhausted troops below.

A drone attack also allows for more oversight than, say, one delivered by an airplane. Bradley Strawser, assistant professor of philosophy at Monterey's Naval Postgraduate School, defended their use to Roy Carroll of the *Guardian*, "Literally every action they [the operators] do is recorded. For a difficult decision they can even wait and bring other people into the room. There's more room for checks and oversights. That to me seems a normative gain." Strawser suggested that using drones is no different from using robots for bomb-disposal teams to reduce the risk to human lives. "It's irrelevant if we use drones, a sniper rifle or a crossbow," he said. The only thing that matters is whether the mission itself is justified.

Given all the advantages of remote warfare, it is little wonder that former CIA Director (and Defense Secretary) Leon Panetta told the Council on International Policy in 2009 that "very frankly, [drone warfare] is the only game in town in terms of confronting or trying to disrupt the al Qaeda leadership."[57]

## The price of push-button warfare

The bigger questions are, however: what are the implications of this new form of warfare? How will it affect us as a society? Or to put it more bluntly, should we really be handing over so much military power to a bunch of geeks?

The reality is that we have no choice. Remote warfare is here to stay. But like any power transition, safeguards must be put in place. Currently, the most important safeguard on the use of drone technology—particularly in the US and UK—is the character of the military hierarchy, many of whom have seen active service in battle, and who are steeped in age-old traditions of courage, sacrifice and honor. War is not theoretical for these people. Their hard-won experiential knowledge shapes their perspective on matters of war, and makes them wiser. They are less susceptible to jingoism or bravado, and more tempered in their responses to provocation—even the more hawkish of them—because they know the cost. They are part of a brotherhood (and sisterhood) forged in real-life combat situations. As a group they exert a powerful influence on the culture of the military, including the employment of new technologies such as drones. But as warfare becomes more remote and automated, our military will inevitably evolve to a point where few, if any, senior people have seen active experience service at the front line. They will all have risen through the ranks of remote warfare systems. These joystick warriors will have a fundamentally different outlook on war—culturally, psychologically and technologically—from the generations of military personnel who preceded them. The smartest of these tech-savvy warriors will rise to the top, just like they have in other professions, such as financial services. These geeks will then call the shots and begin to reshape the military culture in their own image. Indeed, as bizarre as it seems today, we should not be surprised to eventually see four star generals roaming the Pentagon in sneakers and jeans. Everything we know about warfare is about to change.

According to Peter Singer, this transformation is now well under way. He has suggested that the first thing to change is our perception of the "costs" of going to war, for drone technology is so seductive that it creates the illusion that there are no moral or ethical consequences to conducting warfare this way. He says the victims tend to be faceless, and the damage caused by the bombings remains unseen by the general public, because video footage of the attacks is kept under wraps. This creates a fundamental disconnect.

In democracies like ours, there have always been deep bonds between the public and its wars. Citizens have historically participated in decisions to take military action, through their elected representatives, helping to ensure broad support for wars and a willingness to share the costs, both human and economic, of enduring them. In America, our Constitution explicitly divided the president's role as commander in chief in war from Congress's role in declaring war. Yet these links and this division of labor are now under siege as a result of a technology that our founding fathers never could have imagined.

Singer says this technological transformation is affecting the way politicians make decisions about war. "President Obama's decision to send a small, brave Navy Seal team into Pakistan for 40 minutes [to strike Osama bin Laden] was described by one of his advisers as 'the gutsiest call of any president in recent history,'" he wrote in a *New York Times* opinion piece. "Yet few even talk about the decision to carry out more than 300 drone strikes in the very same country. I do not condemn these strikes; I support most of them. What troubles me, though, is how a new technology is short-circuiting the decision-making process for what used to be the most important choice a democracy could make. Something that would have previously been viewed as a war is simply not being treated like a war." Moreover, Singer says, political leaders increasingly act as if they only need to gain approval for operations that send "people into harm's way"—but not for the wars waged "by other means."

Another prominent critic of drone warfare is David Kilcullen, an Australian counterinsurgency warfare expert who advised General David Petraeus in Iraq. He is particularly concerned about the propaganda costs of drone warfare and its capacity to galvanize support for the enemy by causing collateral damage among the local civilian populations. In a study he cowrote for the Center for a New American Security, he put the case that, "Every one of these dead non-combatants represents an alienated family, a new revenge feud, and more recruits for a militant movement that has grown exponentially even as drone strikes have increased."[58]

Similarly, the study's coauthor, Andrew Exum, a former army ranger who has advised General Stanley McChrystal in Afghanistan, told Jane Mayer at the *New Yorker*:

"Neither Kilcullen nor I is a fundamentalist—we're not saying drones are not part of the strategy. But we are saying that right now they are part of the problem. If we use tactics that are killing people's brothers and sons, not to mention their sisters and wives, we can work at cross-purposes with ensuring that the tribal population doesn't side with the militants. Using the Predator is a tactic, not a strategy."

A big part of the problem, explained Exum, is that "there's something about pilotless drones that doesn't strike me as an honorable way of warfare," which is of particular concern in an area like FATA—Pakistan's Federally Administered Tribal Areas. "As a classics major, I have a classical sense of what it means to be a warrior. . . . There's something important about putting your own sons and daughters at risk when you choose to wage war as a nation. We risk losing that flesh-and-blood investment if we go too far down this road."

Which brings us back to the geeks. It would be difficult for anyone—let alone a highly focused number-cruncher—to appreciate the notion of honor in battle, without experiencing it firsthand. Such understanding can only be absorbed through a physical process of osmosis, not just by thinking about it— no matter how bright a person is. It is this holistic, hard-won knowledge about matters of honor and courage that we risk losing when we abdicate the higher echelons of military authority to the new generation of warrior geeks.

As a cautionary example, we only have to observe the banking industry to see what happens when geeks take over, and their sophisticated computer-based models are allowed to run the system. Things blow up.

Of course, it would be naïve in the extreme to propose that the military should try to turn back the clock, and resort to more familiar forms of warfare. Our enemies would love that. But we can take steps to ensure that our next generations of warriors are trained in such a way that they are at least aware of the "human" consequences of their actions. This doesn't necessarily require that they be exposed to real combat—although that would be beneficial. But it does mean that they should be provided with a holistic education that encompasses the cultural and sociological dimensions of their work—and not just how to operate a joy stick. Perhaps they could start, as Andrew Exum did, with an introduction to the classics. Finally, in terms of career advancement, it is important that those personnel with actual combat experience are not sidelined in favor of tech-savvy people who fit the new system. Combat veterans should not be treated as dinosaurs, simply because they cannot manipulate a mouse.

## *The eros geek* (Orgasmus copulus)

The 1970s and 1980s was the golden era of the pornography industry. The studios were jumping, the performers were pumping, and the business was rolling in dough. The people who called the shots were larger-than-life characters, who were often performers-turned-producers, and who knew the ins and outs of the business, so to speak. With few exceptions, they were middle-aged men, many of whom had a penchant for gold chains, strong aftershave and bright red cars from Italy. You can get a glimpse of this testosterone-fuelled environment in the film *Boogie Nights*, starring Burt Reynolds and Mark Wahlberg. It is set in the San Fernando Valley, Los Angeles, the epicenter of the global porn industry.

Fast-forward to today, and the porn industry is unrecognizable compared to what it once was. Many of the original studios have closed, performers struggle to make a living, porn is freely available to anyone with a computer, and video piracy is rampant. It has become a low-margin, high-volume business based on Internet traffic. According to Google's DoubleClick Ad Planner, which monitors web traffic, the largest porn site, XVideos, generates 4.4 billion page views per month, which is three times more than CNN. Another popular site, YouPorn attracts over one hundred million page views every day. Industry estimates suggest that about 13 percent of all web searches are for erotic content. "It is a truth universally acknowledged," wrote Sebastian Anthony of the technology news site Extreme-Tech, "that a person in possession of a fast Internet connection must be in want of some porn."[59]

Given the sweeping changes to the industry, it is perhaps not surprising that the flamboyant Burt Reynolds-style characters of yesteryear have given way to an entirely new generation of porn kings. These are low-key, cerebral types with a flair for numbers and technology. They are hardcore geeks.

## Geeks who gawk

During the first few years of the online revolution, it didn't take much technical know-how to set up a website and provide a video download service. But as the market became more competitive, Internet-based porn businesses realized they had to generate more traffic, which meant optimizing their search results, establishing sophisticated databases, and managing revenue streams from multiple interconnected sites. This was geek work.

Among the first of the new breed to enter the industry were twenty-two-year-old Matt Keezer and two friends from Montreal's Concordia University, Ouissam Youssef and Stephane Manos. In 2003, while still students, they established the site Jugg World—which focused on generously endowed women— and was the first of many sites that would eventually include Brazzers, the biggest porn network on the web. Their story was forensically documented in an article titled "The Geek Kings of Smut" by Benjamin Wallace in *New York*:

"They slept in the office, worked weekends, bought houses near each other in the Montreal suburb of Laval. As their need for manpower exploded, they hired friends, neighbors, classmates—loyalists who could learn on the fly and pitch in as needed. Every year, the company nearly doubled in size. They had 80 employees in 2007, 150 in 2008, 250 in 2009. Youssef was the business visionary, Manos the salesman and motivator, Keezer the savant of search-engine optimization."[60]

It wasn't long, however, before the pay-per-view model of the porn industry—on which Brazzers was based—was threatened by the advent of "tubes" which emulated YouTube by enabling anyone to post porn videos online. With names like YouPorn, RedTube and xTube, these new services offered countless thousands of videos for free. Although many of them were pirated from the established studios, they were protected by the "safe harbor" provision of the US Digital Millennium

Copyright Act, which basically said that the "tubes" were not responsible for "user uploaded" content as they were simply acting as intermediaries.

The next big threat to the Brazzers' business model was the advent of "live cams," such as LiveJasmin.com, which enable users to connect directly with a performer and—for a fee—ask them to perform certain acts. These sites can be set up by anyone with a webcam, a bed and exhibitionist tendencies. They have effectively democratized the industry. As Wallace explained, "If you expand the idea of amateur . . . to encompass a whole new set of outsiders for whom cam sites and tubes have provided a cheap, almost barrier-less way to make, distribute, and sell videos of themselves having sex, well, then, we're living in a grand age of micro-smut, a burgeoning empire of lemonade-stand porn." Wallace suggested that the live cams and tubes have the potential to change the viewer's relationship to erotica itself. "Consumers can also be producers. Posting can be as arousing as watching. We are all porn stars, if we want to be. Maybe porn isn't even really the right word for it anymore, as it evolves from something made to be watched to something made to be shared."

The democratization of the porn industry forced many players, including Brazzers founders—Manos, Youssef and Keezer—to leave the business. They sold their empire to a technological whizz-kid from Germany named Fabian Thylmann, who had been writing software since he was seventeen years old. While in his twenties, he bought a small tube-like site called PrivatAmateure, and by making some small changes was able to double its profits in three months. By the time he was thirty-three, Thylmann controlled one of the largest porn empires in the world, with brands like Brazzers, Mofos, Twisty's,

MyDirtyHobby, Spankwire and four of the ten most popular tube sites.

Despite the erotic nature of his products, Thylmann has insisted that his company is "in essence, a tech company." In a rare keynote speech at Internext, an adult industry conference, he explained, "We're online and have such big sites that we have to be very, very good at the tech side of this business. The online [technical] aspects are the first thing we think about when we do a production. . . . We can always repackage it later and make a DVD or TV segment, but first and foremost it goes online, and that's where it needs to make money."[61]

These comments highlight the fundamental difference between the "geek-driven" porn industry of today, and the old *Boogie Nights* approach. For the business is now less focused on content—such as storylines, talent and background sets—and much more on frequency and turnover. It is about using technology to provide as much variety as possible, to huge numbers of people, in an easily accessible form—and to keep them coming back for more. It is all about frequency and turnover.

The real driver of this new business model, however, is actually a chemical.

## The dopamine connection

When people search for erotic material on the Internet they trigger the brain's "anticipatory" system producing a chemical called dopamine, which is intensely pleasurable. The more we get the more we want. Dopamine is nature's trick to encourage us to find food, seek comfort, and pursue sexual partners, in order to propagate our species.

But here's the thing. The anticipatory—or dopamine—system, is very different from the satisfaction system, because dopamine only creates a craving or desire in us to, say, find the berries in the forest. To become satisfied we have to eat the berries.

The trouble with Internet porn is that it constantly triggers our dopamine (anticipatory) system—without delivering satisfaction. This explains why people will skip through hundreds—even thousands—of images or videos looking for that special one that will deliver the ultimate payoff. But it never quite comes. Kelly McGonigal, a psychologist and author of *The Willpower Instinct*, has explained it this way:

"For much of human history, you weren't going to see a naked person posing seductively for you unless the opportunity for mating was real. Certainly a little motivation to act in this scenario would be smart if you wanted to keep your DNA in the gene pool. Fast forward a few hundred thousand years, and we find ourselves in a world where Internet porn is always available. . . . The instinct to pursue every one of these sexual 'opportunities' is how people end up getting addicted to X-rated websites."

The geeks who run the porn industry understand this addictive phenomenon, which is why they make it so easy to browse countless images (sexual "opportunities") online. In effect, their sites are virtual dopamine dispensers. The more people who visit them, the more revenue they generate from the advertisers on those sites. This addiction-driven business model is tailor-made for an age of instant gratification.

As with any addictive behavior, there are consequences. It is becoming increasingly clear that Internet porn can have a detrimental effect on a person's actual sex life. In a *Playboy* interview in 2010, the musician John Mayer admitted he'd rather masturbate to images than have sex. Pornography is "a new synaptic pathway," he said. "You wake up in the morning, open a thumbnail page, and it leads to a Pandora's box of visuals. There have probably been days when I saw 300 [naked girls] before I got out of bed."[62]

Mayer revealed that:

"Internet pornography has absolutely changed my generation's expectations. How could you be constantly synthesizing an orgasm [with a person] based on dozens of shots? You're looking for the one . . . out of 100 you swear is going to be the one you finish to, and you still don't finish. Twenty seconds ago you thought that photo was the hottest thing you ever saw, but you throw it back and continue your shot hunt and continue to make yourself late for work. How does that not affect the psychology of having a relationship with somebody? It's got to."

Some critics believe that online porn also leads to more extreme forms of pornography, and ultimately, to more bizarre sexual behavior. Gary Wilson and Marnia Robinson, a husband-and-wife team who have researched the effects of pornography for over fifteen years, suggest that by encouraging a constant search for "novelty" (and dopamine), pornography pushes people's sexual tastes to become more extreme. In an article for the Good Men Project organization and magazine, they explained:

"The uniqueness of Internet porn, is that it can goad a user relentlessly, as it possesses all the elements that keep dopamine surging. The excitement of the hunt for the perfect image releases dopamine. Moreover, there's always something new, always something kinkier. Dopamine is released when something is more arousing than anticipated, causing nerve cells to fire like crazy. In contrast, sex with your spouse is not always better than expected. Nor does it offer endless variety. This can cause problems because a primitive part of your brain assumes quantity of dopamine equals value of activity, even when it doesn't. Indeed, porn's dopamine fireworks can produce a drug-like high that is more compelling than sex with a familiar mate."

Over time, suggest Wilson and Robinson, "too much stimulation can numb the pleasure response of the brain for a time, pumping up cravings for more novel stimuli."[63]

This view has been challenged by two neuroscientists, Ogi Ogas and Sai Gaddam, who studied the sexual behavior of more than a hundred million men around the world by observing what they do within the anonymity of the Internet. Their analysis, which culminated in their book *A Billion Wicked Thoughts*, suggests that although the Internet does change sexual tastes, most men search for the same erotic content over and over again. In an article in *Psychology Today*, Ogas wrote:

"The majority of people who searched for sexual material on AOL searched for fewer than four different (sexual) interests. Less than 1% of users who searched for sexual content searched for more than eight interests. In fact, there is no evidence that viewing porn activates some kind of neural mechanism leading one down a slippery slope of seeking more and more deviant material, and plenty of evidence suggesting that adult men's sexual interests are stable. . . .None of [the] data suggests that users steadily escalate the 'hardcore-ness' of the material they seek. Men purchase subscriptions to the same type of fetish website over and over again. When men visit tube sites, they view the same genres of videos over and over again.

Men tag the same kind of videos again and again, and read the same kind of stories again and again."

Perhaps the most surprising finding of the research was that it was women, rather than men, whose sexual tastes were more likely to change over time. "Women's searches for porn roam all over," noted Ogas. "Their erotica is far more complex, diverse, and demanding than men's. The most popular erotic artifact for women is the romance novel in all its variants, such as *Fifty Shades of Grey*—stimuli that requires a significant investment of time and mental effort to consume, then leads to many more hours of discussion and analysis with other fans. In contrast, the most popular erotic artifact for men is the 90-second video clip."

Yet despite these conflicting views on how Internet porn influences us, there is a broad consensus that it is changing the way we, as a society, view sexuality. This pornification process is being driven largely by the industry geeks, who have effectively turned a large portion of the Anthropocene into a red light district—one that is always open, easily accessible, and floods our collective nervous system with dopamine.

It would be a mistake to believe, however, that this powerful dopamine effect is confined to pornography. For it permeates, albeit to a lesser extent, the entire online world. Millions of people now live in a constant state of salivary anticipation of the next Facebook update, tweet, SMS or email—all of which trigger a dopamine surge. We often don't recognize how addicted we have become to the online world until, say, our iPhone breaks down and we suffer withdrawal symptoms. We feel cut off.

Meanwhile, the architects of this digital reward system, the geeks, are busily inventing more effective ways to keep us plugged in. Emerging new technologies such as immersive 3-D and tactile feedback will further blur the line between reality and fantasy. Like so many other dimensions of our lives in the Anthropocene, sex will become increasingly virtualized.

We have certainly come a long way since the days when, in 1819, King Francis I of Naples took his wife and daughter to visit the Pompeii exhibition at the Naples National Archaelogical Museum, and was so embarrassed by the erotic images that he ordered the artwork be locked away in a secret cabinet where it could only be viewed by "people of mature age and respected morals." Those same images can now be viewed online by millions of people at the click of a button.


## *The political geek* (Vox technocratia)

On February 17, 2011, President Obama had a private dinner with a small group of people in a house in Woodside, California. The gathering of mostly middle-aged men looked like just another "meet-and-greet" for a President keen to enhance his pro-business credentials in an election year. But it was more than that. For these were no ordinary executives. They were the CEOs of some of the most powerful technology firms in the world—Steve Jobs of Apple, Eric Schmidt of Google, Larry Ellison of Oracle, Mark Zuckerberg of Facebook, Dick Costolo of Twitter, Reed Hastings of Netflix, John Chambers of Cisco, and Carol Bartz of Yahoo.[64]

Obama had flown from Washington to meet with these super geeks to discuss his proposals "to invest in research and development and expand incentives for companies to grow and hire," explained Jay Carney, press secretary of the White House.

This dinner was also significant for another reason. The attendees represented a changing of the guard in terms of power and influence in America. In previous decades, it was customary for an American president to meet privately with the titans of the manufacturing industries or the heads of the big media conglomerates. Now Obama was meeting with the gatekeepers of the information economy. These new "masters of the universe" also controlled something of vital interest to any politician—social media.

The rise of Twitter, Facebook and other social networks, has transformed the political equation in America and around the world—by enabling people to communicate, connect and empower each other to action. In the Middle East, for example, social media helped to topple the governments of Tunisia, Libya and Egypt. The porous and interconnected nature of social media makes it difficult to stop the momentum once it gains traction, as the Egyptian government discovered when it tried to shut down the local Internet. Google promptly launched a new "speak to tweet" service to allow people in Egypt to send Twitter messages by dialing a phone number and leaving a voicemail. Digital information flows like water, constantly seeking new channels and outlets.

The fluid nature of social media has forced politicians to fundamentally rethink their approach to communications. The "top-down" approach, whereby messages are delivered through a hierarchy of "old media," such as television, radio and newspapers, no longer works in the new "horizontal" environment, where messages spread virally and obey the laws of epidemiology. To deal with this new environment, many politicians now employ tech-savvy geeks to manage their online activities and shape voter perceptions.

The first presidential candidate to capitalize effectively on digital technologies was Barack Obama during his 2008 campaign. He needed a technological edge because the odds were so heavily stacked against him. At the time, Obama was a relatively unknown African American with an Islamic-sounding middle name—Hussein—who was running for president in a majority-white country that had never elected a black leader. He faced not only the barrier of racism; his mixed cultural background also made him seem like an outsider at a time when the American public had become increasingly xenophobic, following years of external terrorist threats. Consequently, he was confronted with formidable opposition, both from within and outside his own political party.

Obama was also acutely aware of what happened to the previous Democratic presidential contender, John Kerry, in 2004. Although Kerry seemed to have everything going for him—he was tall, wealthy and a decorated war hero—his campaign was derailed by a viral Internet campaign launched by some veterans who had served with him on the Swift boats in the Vietnam War. They attacked Kerry's fitness to be commander in chief, and by the time he responded, it was too late.

To avoid a similar fate during his 2008 campaign, Obama's team established a website called "Fight the Smears," which enabled them and their supporters to react immediately to any threat. They developed social media programs that raised hundreds of millions of dollars, which, in turn, funded expensive television campaigns to drive home Obama's mantra of "Change." The net-savvy campaign team operated like an Internet start-up and spread their message with viral intensity.

The lesson from all this was clear—digital communication was no longer just an adjunct to electioneering; it had become the heart of the process. The age of Internet campaigning was born.

When it came to his 2012 reelection strategy, Obama took no chances. This time he was up against a particularly well-funded candidate in Mitt Romney, the Governor of Massachusetts, who had the support of the big business community, and super wealthy donors such as the Koch brothers. "This will be the geeks versus the billionaires election," quipped Sam Graham-Felsen, Obama's chief blogger in the previous campaign.[65]

The geek that Obama chose to run his campaign was a forty-two-year-old whiz kid named Jim Messina. In an interview in *BusinessWeek*, Messina recalled how Obama approached him for the role while they were in the surf together at Waikiki Beach in Hawaii. "I've got a favor I want to ask," Obama said. "I'd like you to run the reelect."

Messina replied: "You have to understand, this will be nothing like the last campaign."

Obama: "I thought the last one went pretty well."

Messina replied: "It did. But everything is different now."[66]

Messina was referring to the way technology had—once again—totally transformed the political landscape. Campaigning had moved beyond creating a nice website and viral Internet programs. It was now all about slicing, dicing and analyzing the electorate to an infinitesimally precise degree, and then using social media to drive home the message. Voting intention models had become as sophisticated as NASA trajectory calculations. Political intuition was out, multivariate analysis was in. There were new weapons in the online arsenal too—not just Facebook and Twitter—but also Google's new "network blasts," which enable a politician to saturate a particular region with advertisements on certain days. So if you lived in, say, Ohio in the United States, or Manchester in the United Kingdom, and logged on to Google, you would see ads from a relevant candidate all day long.

According to Elliott Suthers, writing in *Forbes*, "politics used to be about more than percentage points and message penetration. It used to be about guts, about connecting with voters, shaking their hands and getting to know the issues that really mattered to them. Now, candidates are propped up by a cadre of iPad toting math geeks, who have at their fingertips more information than a regular field director could ever dream of."[67]

Shortly after Messina accepted Obama's job offer, he sought out the best brains in the business, including Apple's Steve Jobs and Google's Eric Schmidt. Messina told *BusinessWeek* that when he showed Jobs the proposed campaign strategy, the Apple guru immediately picked holes in it, and suggested ways to improve their reach and message to voters. "He knew exactly where everything was going," Messina recalled. "He explained viral content and how our stuff could break out, how it had to be interesting and clean."

Messina and his team took over an entire floor of the Prudential Building in Chicago for nine months to devise a comprehensive digital strategy from the ground up. They built proprietary databases, reporting and communication systems, in order to target swinging-voters. They applied their sophisticated analytical and technological skills to leveraging social media. They created a QuickDonate program that made contributing to the campaign simple. Most importantly, they built a campaign machine that could react as fast as lightning, as was demonstrated by an incident later reported on the Politico website.[68]

At 12:23 p.m. on August 31, 2012, Mitt Romney told a home crowd, "No one's ever asked to see my birth certificate." This comment immediately associated Romney with the extreme fringe of his party that believed Obama was some sort of Manchurian candidate from Kenya. Within sixty seconds his words were tweeted by a *Washington Post* reporter, and four minutes later were online. By 12:41 p.m. Romney's ill-advised comment was on YouTube, and three minutes later Obama's team issued a statement denouncing Romney's words. Previously, a "gaffe" like this would have taken a full day to percolate through the media channels, but the nature of online media has compressed the twenty-four-hour news cycle into just twenty-four minutes.

Over the ensuing months the campaign was bitterly fought, with both candidates bruised and battered. In the end, however, it was Obama who emerged victorious, by winning not just a healthy majority of Electoral College votes—which determine American election outcomes—but also the popular vote. The geeks had won.

Interestingly, the outcome of the election had been predicted well in advance by a number-crunching whiz kid named Nate Silver, a thirty-four-year-old statistician and former economist. He first made his mark designing a computer model that could forecast the performance of baseball players, and in 2007 he turned his talents to political forecasting. Traditional pundits, who based their forecasts on old style polling and a large dose of gut feeling, were highly skeptical of Silver's work—especially those on the conservative side who saw him as politically biased. The feeling was mutual. As British author Salman Rushdie observed, "Silver's contention, that TV pundits are generally no more accurate than a coin toss, must now be given wider credence," because Silver offers the "consistently best analysis of US election(s)."[69]

In fact, Silver's election forecast was so spot-on that he correctly predicted the result in all fifty states. This prompted Rachel Maddow, who hosted MSNBC's election coverage, to suggest, "You know who won the election tonight? Nate Silver."[70] His combination of sophisticated mathematical models and algorithms has changed the face of political forecasting in the same way Messina and his team reinvented the campaign process.[71]

## Wonks who woo

The growing influence of geeks on the political process is understandable, given the pivotal role that technology now plays. What is surprising, however, is that many of them are now stepping out of the backrooms into the political spotlight. This is despite the fact that geeks have never been known for their political skills, given their tendency to be introverted, obsessive, and, well, let's be frank, a bit weird. Whereas politicians tend to be "people's people," who love nothing more than getting out amongst the crowds—feeding on the adulation. Unlike the typical geek, the best politicians are highly empathetic and sensitive to the public mood—able to detect infinitesimally small variations in voter sentiment.

Notwithstanding their formidable seductive talents, however, politicians are not always the brightest people, nor the most intellectually curious. The very skills that enable them to be so persuasive can convince them that they are right, regardless of the facts. They tend to prefer ideology to evidence. Which explains why politicians will often jump on a popular bandwagon, despite overwhelming evidence against it. Recently, for example, politicians in many countries joined in a fear campaign waged against MMR (measles, mumps and rubella) vaccines for children, despite the incontrovertible evidence in support of their use.

The problem is, according to Mark Henderson, author of *The Geek Manifesto: Why Science Matters*, is that "What science admires as intellectual honesty is seen in Whitehall as the stuff of the gaffe."[72] Scientists are used to changing their minds when confronted with new evidence, whereas for politicians this can be seen as a lack of conviction. So they stick to their guns in order to appear consistent. Henderson says this attitude is unfortunate because, "A more scientific approach to problem-solving is applicable to a surprisingly wide range of political issues, and that ignoring it disadvantages us all."

Geeks know this, which explains why, as in so many other fields, they are emerging from the shadows into the political limelight. For example, during the 2010 contest to lead the British Labour Party, the four main contenders were all former policy advisers, or wonks—Ed Balls, Andy Burnham and the Miliband brothers David and Ed. The Milibands are precociously bright and Oxbridge educated; David is a former head of the Downing Street Policy Unit, where he was nicknamed "Brains" by his colleagues. Ed Balls had the dubious distinction of having persuaded the former Chancellor Gordon Brown to use the term "post-endogenous growth theory" in a speech on the economy. The only woman contender for the Labour leadership, Diane Abbott, described her competition as "geeks in suits." They were the kind of people Hermann Hesse immortalized in his classic book *The Glass Bead Game*, in which the most intellectually gifted battle it out for the title of "Magister Ludi" (master of the game). Eventually, after much factional infighting, it was Ed Miliband—arguably the geekiest of them all—who became leader of the Labour Party. The wonk won.

Over the next few years, Miliband found it tough going, especially during his weekly appearances in the "bear pit" of the House of Commons, where he faced off against the devastatingly smooth David Cameron. But over time, Miliband grew into the role and inflicted real damage on Cameron's leadership credentials. Indeed, by late 2012, he was widely lauded by his own party and the media. Tom Peck of the *Independent* newspaper was so impressed by Miliband's performance at the Labour Party's annual conference that he wrote, on October 6, 2012:

"Downtrodden geeks of the world, unite! You have nothing to lose but your adenoids. As well as a much-needed rallying call to the Labour Party faithful, Ed Miliband's surprisingly triumphant conference speech this week provided final confirmation of a truth that has been slowly but surely emerging: there has never been a better time to be decidedly uncool . . . it is fitting that it has taken a bona fide, genuine geek to bring [uncoolness] surging into the mainstream. . . ."

Mr. Miliband told the BBC's Andrew Marr he was "proud" to be thought of as a pointy-headed policy wonk (the kind of pejoration only policy wonks understand). It is not so long ago he confessed to Piers Morgan that his main talent at school was being "really good at Rubik's cube." And yet, suddenly, it is more than just Ed's brother and special advisers who are having a "hang on a minute" moment, and thinking that this slightly awkward, slightly odd chap might just end up as Prime Minister. Whatever Mr. Miliband thinks One Nation means, in the Benjamin Disraeli phrase he repeated hundreds of times, come 2015 this one nation might just have a geek running the show.[73]

It seems that the geek politician has arrived.

Another geek-turned-politician success story is the former foreign minister of Australia, Bob Carr. Before entering politics he was a journalist, intellectual, author of several books, and an authority on US Civil War history. He frequently moved in literary circles — unusual for a politician — and counted the late Gore Vidal as a personal friend. Despite his overt intellectualism — or perhaps because of it — Carr has managed to make substantial and enduring contributions to Australian politics at both the state and federal levels, first as premier of New South Wales and then as the nation's foreign minister.

Even in the United States, the birthplace of shallow "image politics," geeks are making inroads into the political process. In 2012, the Republican presidential nominee, Mitt Romney, chose as his prospective vice president a forty-one-year-old economics guru named Paul Ryan, who is renowned for his command of fiscal details. Apart from the intellectual clout he brought to the Romney ticket, Ryan's geekiness made him seem less vulnerable to accusations of "sleaze" — a critical factor in US politics — especially since the 2004 Democrat vice presidential candidate, John Edwards, fathered a child with a campaign worker. A geek like Ryan, it was believed, would probably be too busy solving problems to get into mischief.

Political geeks are not an entirely new phenomenon, of course. America's founding fathers included Benjamin Franklin, a scientist, musician, inventor, author and diplomat; and Thomas Jefferson, a polymath who spoke five languages, and was deeply interested in science, philosophy, and religion. Britain's Sir Isaac Newton (1642–1727) was an elected member of parliament for Cambridge University, although it is reputed that the only time he spoke during a parliamentary session was to ask for a window to be closed because he could feel a draught.[74] More recently, we have seen people like Henry Kissinger and Condoleezza Rice, who were able to apply their formidable cognitive talents to the pragmatic demands of high office. Jimmy Carter too, the thirty-ninth President, was a science graduate with an obsessive focus on detail. He revealed his true colors when he admitted diffidently to *Playboy* in November 1976 that "I've looked on a lot of women with lust." Only a geek would talk about sex like that.[75]

Notwithstanding these exceptions to the rule, most intellectually inclined candidates have tended to hide their inner geek behind a veil of "folksiness" so that voters can more easily relate to them. But this is now changing. As the world becomes more complex, people expect their elected representatives to be able to deal with the intellectual challenges of the Anthropocene. As in so many other dimensions of our lives, in the political arena we are becoming more tolerant—and even respectful—of geekiness.

## The clickable candidate

The rise of the geek politician is being accelerated by the "virtualization" of the democratic process, which ensures that most of us never see our candidates in the flesh anymore—only on a computer screen or TV. This process has been going on for some time, of course, ever since John Kennedy used his television presence to beat Richard Nixon in 1960. But the Internet has taken it to a whole new level by enabling candidates to create the illusion that they are connecting with us—via social media and so on—without ever doing so in the flesh. This lack of physicality means we don't get an opportunity to register the caliber of the politician at the visceral—or gut—level. We only see their screen persona, which is a carefully crafted, and ultimately distorted, one. That is, a facsimile.

This virtualization benefits aspiring geek politicians because they feel more at home in the online world. It is the water they swim in. They don't need to rally town hall meetings with soaring rhetoric, or make instant connections with strangers on the streets. They can connect with voters through technology, on their own terms. Over time, this evolution means that we will see less of the old style evangelistic "Elmer Gantry" politicians, and more of the cerebral "Magister Ludi" types—the David Milibands of the world.

But not just yet, because there is one last hurdle to overcome before the democratic process is completely virtualized. This is the voting process itself, which remains, in most countries an experiential, tangibly physical process. That is, we physically turn up to the voting booths on election day, brush past the campaign workers thrusting how-to-vote-cards at us, wait in line for our turn, enter into the booth and use a pencil to mark our preferences on a piece of paper. Along the way, we experience a cacophony of sights, sounds, smells, sensations and moods—the typical hubble-bubble of election day. It is only a matter of time, however, before this last remaining political ritual is digitized, and we cast our votes online—by clicking a button, thereby turning the democratic process into a political version of *America's Got Talent*. The Canadian philosopher and scholar Marshall McLuhan once predicted that "[p]olitics will eventually be replaced by imagery. The politician will be only too happy to abdicate in favour of his image, because the image will be much more powerful than he could ever be."[76]

This abdication is now well under way.

# The gamer (Genghis cyberis)

Lim Yo-Hwan is an unusual superstar, at least by Western standards. He is not an actor, singer, football player or even a chef. This boyish geek from South Korea is a video game player—perhaps the best the world has ever seen. The BBC's Lucy Ash visited him in his dressing room in 2008 and described the scene. "[The room] is filled with presents from his admirers, such as chocolates, snacks and hot pads to keep his precious fingers warm." Outside a gaggle of girls was waiting for their hero to appear, hoping for his autograph. When Ash asked a girl in pink fluffy slippers about a roll of paper in her hand, she glanced furtively at her boyfriend and unrolled it to reveal a poster decorated with hearts reading: 'Lim Yo Hwan—Marry Me!'"[77]

To Lim's legions of followers around the world he is known simply as the Terran Emperor. His game of choice is StarCraft, a science-fiction strategy game set in the twenty-sixth century that is widely regarded as one of the most influential video games of all time. Lim rose to dominance by playing aggressively, and turning around dire situations when all seemed lost.

In Lim's home country of South Korea, where gaming is a national obsession, nearly half the fifty million population play online games. Many of them head for the thousands of dim, smoky Internet cafés known as PC bangs, where they immerse themselves in life-or-death battles between soldiers, zombies, aliens, car thieves, or any number of virtual protagonists. Many more people tune in to the live TV broadcasts of professional league games, which are sponsored by big corporations, such as Samsung. The biggest arena of all is the World Cyber Games, the Olympics of the gaming world, which is often won by a South Korean.

It is not just professionals such as Lim who spend many of their waking hours playing games. Millions of people around the world do—and they are not all young either. The average age of a gamer in the United States is thirty-four years old, and in European countries it is slightly older. Also, over 40 percent of gamers are now female—many of whom started by playing Angry Birds and Bejeweled on their smartphones, before graduating to more complex games. Indeed, gaming is now such a huge industry that it eclipses cinema as the largest sector of the global entertainment business—with annual revenues topping $70 billion in 2012. When Call of Duty: Modern Warfare 3 was released in late 2011, it generated sales of over $1 billion in just sixteen days—a better result than was achieved by *Avatar*, arguably the most successful movie ever made.[78]

## People of the screen

The world of video gaming has certainly come a long way since I first played Space Invaders—a classic shoot 'em up game—during the early 1980s. It was a relatively small industry then, and there were no professionals, just amateurs like myself. There were also very few hardcore players—or gamer geeks—who would practically live for the game. For most people, it was an intermittent hobby, and it would have been rare to play for more than an hour or two a day, if that. The biggest difference, however, between those early days and now, is that video games used to exist in a dimension of their own. That is, there were few overlaps between the fantasy world of the games and the real world outside.

It is fundamentally different today. The rise of virtual technologies has transformed us into "people of the screen," who

THE RISE OF THE GEEK

navigate reality with our iPads or smartphones, much like in a video game. Indeed, we are all gamers now, to some degree. We saw earlier how, for example, financial traders are actually high-stakes video gamers, who, incidentally, can inflict far more damage than Lim can playing StarCraft. We also saw how joystick warriors, sitting before computer screens, are controlling drones from thousands of kilometers away. Even commanders of tanks and other military hardware now use controls that look remarkably similar to PlayStation game consoles. Conversely, video game companies try to make their programs as realistic as possible by imitating real-life military specifications and capabilities.

According to Alex Rayner, writing in the *Guardian* newspaper, the level of realism in today's games is extraordinary.[79] He described his experience playing a combat game which enabled him to join the 2011 raid on Osama bin Laden's compound in Pakistan: "It's Monday night, the kids are in bed, and I am trying to kill Osama bin Laden. I stalk through his Abbottabad compound and I aim my rifle at the first person I see, only to discover he's my brother in arms, aka 'OverdoseRocks.' So I walk downstairs into a prayer room, at which point my gun accidentally goes off. Then the mission is over. We were victorious."

Rayner then joins a team of US soldiers in the 2007 surge in Iraq, and battles his way along a palm-lined boulevard, dodging strafing bullets, before being killed in an ambush. In his dying moments, he is offered the opportunity to play another game in which he can become a cheetah and kill an antelope, but he declines. "I have had enough bloodshed for one evening," he wrote.

The point is that video games are becoming so realistic that they blur the line between actuality and fantasy. They can also be intellectually demanding, especially the strategy-style games. This makes them particularly appealing to the type of personality that is obsessive by nature, and is good at dealing with lots of rapidly changing data. That is, geeks.

The world of gaming is a natural habitat for these people because it offers them an opportunity to compete against each other in a virtual world, in which the rules of social engagement are more flexible. In this stigma-free environment, nonneurotypical people feel more at home.

Stephanie Bendixsen, who cohosts—with Steven O'Donnell—*Good Game Spawn Point*, a popular ABC TV show about gaming, has said: "We receive a lot of positive feedback from parents of children with autism or learning disabilities, who find that the games provide a great environment for their kids to interact socially. They often don't have these opportunities in their real lives, but online they can engage with others, explore and be mentally stimulated in a social environment."[80]

The sheer complexity of some games has also spawned a huge compendium of knowledge for gaming geeks to grapple with and exercise their minds on. "The level of detail that many gamers pay attention to is staggering," said Mike Langlois, a psychotherapist who teaches at Boston College, "whether it be leveling a profession to 525 in WoW [World of Warcraft], unlocking every achievement in Halo 3, or mapping out every detail in the EVE universe. . . . This is meticulousness." The information compendium for Warcraft, for example, now comprises nearly 100,000 pages of information, and has its own wiki online. According to the founder of Wikipedia, Jimmy Wales, these types of wiki sites are where you see "communities of passionate fans come together."[81]

Some gamers, however, take their passion to the extreme. They spend countless hours in front of their screens, which can exact a heavy toll—physically and mentally. In July 2012, a Taiwanese teenager named Chuang died in an Internet café after playing Diablo III for forty hours straight, with no food or sleep. Others have died in similar circumstances; sometimes in their own homes, behind locked bedroom doors. Even a player like Lim—a virtual immortal—was forced to take a break because of concerns about his shoulder and his general well-being. Gamers often push themselves to the limits of their physical, intellectual and emotional capabilities. The addictive nature of video games is a concern for parents, who see their children mesmerized for hours on end before a screen. They worry about the effect this behavior will have on their development. For those of us who grew up in a time and place where kids were more outdoors-oriented, it doesn't seem right that children spend so much time inside on their computer games—even in this era of heightened awareness of "stranger danger."

But are we jumping at virtual shadows? It seems that every generation of parents worries about novel influences on their children. Even Plato in *The Republic* argued that poetry and plays should be restricted because of their dangerous effects on the young. Are we conveniently forgetting how much of our own childhood was spent watching TV cartoons or sitcoms— those endless repeats of *Gilligan's Island*, *Happy Days*, *Neighbors* and *The Simpsons*. Should we really be so concerned about this next generation of entertainment that has captured the hearts and minds of our children?

Bendixsen thinks not. "If I was a parent, I would much rather see my children playing a game where they have to solve complex problems with their friends than have them sitting around passively watching hours of TV. Gaming is a far more interactive and stimulating medium."[82] Hard fun, not easy fun.

Dr. Peter Gray, a research professor of psychology at Boston College who focuses on educational psychology, has suggested—like Bendixsen—that we should not worry too much about video games, and that they can actually be beneficial to children. In his recent book *Free to Learn* and in an article in *Psychology Today*, he argued that because computers are the most important tools of modern society: "why would we limit kids' opportunities to play with them? Our limiting kids' computer time would be like hunter-gatherer adults limiting their kids' bow-and-arrow time. Children come into the world designed to look around and figure out what they need to know in order to make it in the culture into which they are born. They are much better at that than adults are. That's why they learn language so quickly and learn about the real world around them so much faster than adults do."[83]

He said that repeated experiments have shown that playing fast-paced action video games can markedly increase players' scores on tests of visual-spatial ability, including tests that are used as components of standard IQ tests. Other studies suggest that, depending on the type of game, video games can also increase scores on measures of working memory (the ability to hold several items of information in the mind at once), critical thinking, and problem solving. Gray said there is also growing evidence that kids who previously showed little interest in reading and writing are now acquiring advanced literacy skills through the text-based communication in online video games.

In the context of unnatural selection, it could be that video games are another environmental stimulant pushing our society to become smarter and more cognitively dextrous. They

are an integral part of the Anthropocene's development process. Perhaps, at an instinctual level, young people know this,
and turn toward video games to train them for a high-tech
environment. These games could act as life skills simulators.
Indeed, this notion may not be as far-fetched as it sounds. Gray
has suggested that (massively) multiplayer online role-playing
games (MMORPGs) can be extraordinarily sophisticated and
challenging, and can accelerate a young person's intellectual
and social development. Gray cited the example of World
of Warcraft, wherein players create a character (an avatar),
through which they enter a complex and exciting virtual world
that is occupied by countless other players, who in their real-
life forms may be sitting anywhere on the planet. "Players go
on quests within this virtual world," Gray explained, and along
the way they meet other players, who might become friends or
foes. Players may start off playing solo, avoiding others, but to
advance to the higher levels they have to make friends and join
with others in mutual quests. Making friends within the game
requires essentially the same skills as making friends in the real
world. You can't be rude. You have to understand the etiquette
of the culture you are in and abide by that etiquette. You have
to learn about the goals of a potential friend and help that individual to achieve those goals. Depending on how you behave,
players may put you on their friends list or their ignore list, and
they may communicate positive or negative information about
you to other players. The games offer players endless opportunities to experiment with different personalities and ways of
behaving, in a fantasy world where there are no real-life consequences for failing.

Because socio-dramatic online games like Warcraft are very
much anchored in an understanding of the real world, they can

teach valuable lessons about human behavior. This is borne out by a recent study commissioned by the IBM Corporation, which concluded that the leadership skills exercised within MMORPGs are essentially the same as those required to run a modern company.

It is becoming increasingly clear that video games can foster our cognitive and behavioral development. Some critics worry, however, that the violent nature of certain games can trigger negative behavior. They point to examples of people who have gone on killing sprees and were subsequently found to have been playing violent games. This would suggest a "monkey see, monkey do" phenomenon, whereby gamers imitate what they see played out on a computer screen.

Yet, a number of recent studies suggest there is no correlation, and that, paradoxically, in some instances aggressive video games can actually reduce such behavior. When the US Supreme Court recently adjudicated on the issue, after much testimony by numerous parties, it concluded, "Studies purporting to show a connection between exposure to violent video games and harmful effects on children do not prove that such exposure causes minors to act aggressively." The Australian government reached a similar conclusion in 2010, when it was faced with an onslaught of petitions trying to ban or restrict violent video games. It eventually introduced legislation that allowed such games, but under a restricted R 18+ guideline.

It turns out that the most important factor affecting a child's propensity for violence is not video games but the behavior of the adults around them. A study by Christopher Ferguson and his colleagues at Texas A&M International University tracked one-hundred sixty-five young people over a three-year period and assessed their video-game play and various other aspects of their lives. They found no relationship at all between exposure to violent video games and real-world violence committed by these young people. They did find, however, that kids whose parents or friends were violent were significantly more likely to engage in real violence themselves than were kids whose parents and friends were not violent.

These findings are reassuring. They suggest that, currently, gamers make a clear distinction between their actions online and in the real world. So just because a person acts like Genghis Khan or Attila the Hun in a game doesn't mean they will do the same in their workplace or the local shopping mall.

The trouble is, however, that, as discussed throughout this book, the demarcation lines between reality and virtuality are becoming increasingly blurred. In fact, in some industries such as financial services, they have become as one—which is why we do see real-life Genghis Kahn behavior in our financial markets, with dire consequences. We are living in an age of porous perceptual boundaries in which it is increasingly difficult to perceive differences between substance and reality. This evolution is not just being driven by game technology, but also by game aesthetics. Video games are now recognized as a legitimate art form which—like any art—exerts a subtle, yet powerful, influence on societal tastes, aspirations and morals. In March 2013, the New York Museum of Modern Art opened its first exhibit of video games, featuring classic games such as Tetris, Pac-Man and The Sims, inside the same building that houses works by Claude Monet, Vincent Van Gogh, and Frida Kahlo. Tom Bissell, a journalist and lifelong gamer, who wrote the book *Extra Lives: Why Video Games Matter*, noted:

"Around 2006, 2007, a handful of games started coming out that, as someone who played games, but didn't think of them as like a viable artistic medium, made me think, 'Wow things have gotten extremely compelling formally.' I mostly associated video game storytelling with unforgivable clumsiness, irredeemable incompetence, and suddenly I was finding the aesthetic and formal concerns I'd always associated with fiction: storytelling, form, the medium, character. That kind of shocked me. The graphics, storytelling and interactivity of gaming have all made tremendous leaps forward in recent years."[84]

Over time, the increasing influence of video games—artistically, technologically and behaviorally—is pushing our society to become more like a giant 3-D multiplayer video game, in which we ourselves will evolve into gamer geeks. To get a glimpse of this future, we need to look no further than our children.

## *The late bloomer* (Geekus maturum)

Until his mid-forties, Max Fulcher had not been a fan of technology, even of the most rudimentary kind. Although he was a successful advertising man—driven by deadlines—he didn't wear a watch, owned no gadgets, avoided using photocopiers and fax machines, and preferred on/off switches to intricate remote controls. He liked things he could touch and feel. Instead of a CD music player he installed a baby grand piano in his house. And he rarely, if ever, peeked below the bonnet of his big black Mercedes car. Fulcher wasn't exactly a Luddite, but he came close.

All this changed in the 1980s when he sold his advertising business to travel Asia, hunt down orchids, and illustrate them. He had been a keen sketcher since he was a child, and even when he became CEO of his company he often conceived the advertising layouts himself.

Despite his aversion to computers, a tech-savvy friend showed him how to use an Apple machine to draw pictures freehand using a mouse. "I was skeptical at first," he told me, "given my track record on such things. But I could see that the mouse was really a kind of pen or paintbrush that I could use physically—so I could relate to it." He took the plunge and bought an Apple Mac desktop computer with Adobe Illustrator already installed. After much trial and error he learned to master the software and translate his drawing skill onto the Mac. "I learned by doing, as I don't read manuals."[85]

Over the next two decades, with trusty mouse in hand, he drew thousands of orchids and became a top orchid illustrator. He also wrote stories, took countless photos, produced popular blogs, and published his first book, *The Enchanted Orchid*. The ad-man reinvented himself.

"I see myself as a late bloomer," Fulcher said. "The passion for drawing was always there, but the computer showed me the pathway. I have Steve Jobs to thank for that, because he made things easy to use."

It turns out that Fulcher may have had one other advantage. When he turned seventy he was diagnosed by his doctor as having ADHD. "The minute I heard him say those four letters," Fulcher recalled, "so much of my life fell into place. I have always been this hyperactive, full-on energetic person, whose attention snaps to the next thing. I was never a big reader, but give me a picture and I get it immediately."

As noted earlier, people with ADHD are sometimes blessed with remarkable visual skills. They can often pick patterns, see trends, spot anomalies and get the gist of a picture much faster than a so-called normal person. They live in a visual world. Fulcher's diagnosis may help explain why—despite him being a technophobe—he was able to make such a successful transition to the digital world. Like many people we have met in this book, he had a dormant, innate trait that resonated with the technocentric environment of the Anthropocene. So instead of retiring from work, as many people in his position would have done, he was able to embark on a new career and leverage his ADHD in ways that would not have been possible in the old environment.

"The odd thing is," Fulcher said, "I once promised myself I would never own a computer, and that I would resist all that tech stuff. But here I sit at seventy-seven working on my first e-book. Another challenge."

Max Fulcher's story highlights one of the most intriguing phenomena of the digital age—the rise of the late bloomer; people who have discovered the liberating power of technology at a relatively older age and are determined to make the most of it.

They may be grandparents using Facebook to keep up with grandchildren, or retirees embarking on an online course at a university, or mid-career professionals who suddenly decide to design apps for a living. They may be people who, years earlier, deferred their dreams while occupied with the three Ms— money, mortgage, and motherhood (or fatherhood).

Their first brush with digital technology might have been a simple gadget like an iPod music player, or perhaps an e-book reader. Those with fading eyesight may have discovered how much easier it is to read a book on a Kindle, on which you can increase the size of the typeface. According to research conducted in 2011 by a team at Johannes Gutenberg University of Mainz in Germany, older people read faster on an iPad than they do reading "real" books. In fact, more than one in ten (11 percent) of seniors over sixty-five years of age in the United States now owns an e-reader, up from 3 percent in 2010.[86]

In addition to making it easier to consume information, digital technology encourages older people to be more productive by making the creative process more fluid—any mistakes can be easily corrected, unlike committing a work to paper or a fixed medium. There is less fear of messing up. So, for example, a mid-career changer like Max Fulcher can launch himself into drawing knowing that his efforts are works in progress. When he makes a mistake, it can be easily fixed because the medium is so inherently flexible and transient. The new technology takes some of the pressure off by making drawing a more playful process, so he can dabble and express himself freely.

This is not to suggest, of course, that all mature people need technology to succeed. Long before the digital age came along, there were late bloomers in every sphere, some of whom achieved great fame. The architect Frank Lloyd Wright, for example, didn't undertake his significant work until after the age of sixty-five; artist Paul Cézanne had his first one-man show at the age of fifty-six; Alfred Hitchcock made his most famous films after he turned fifty. Katharine Graham, the former *Washington Post* publisher, only hit her stride once she was well into her fifties and steered her publication through the Watergate scandal.

But such people were truly exceptional—giants in their fields—who would probably have succeeded in any environment. What the digital age does is make it easier for more people to become late bloomers and to unleash their talents, in the same way that it can help non-neurotypical people to turn their weaknesses into strengths.

Importantly, the digital age also makes it easier for late bloomers to gain recognition for their work, which, for a creative person or innovator, can be crucial. As the philosopher Alain de Botton, author of Status Anxiety, wrote, "Nobody wants to be a nobody." Now anyone can build a profile by creating a website, or by using Facebook or Twitter, whereas in the old days if someone was considered to be past their prime, it would be unlikely for them to receive attention from the mainstream media, or even their peers. Unless they were a remarkable—or notorious—figure, they would remain in the shadows. This ageist bias has been particularly pronounced in industries such as music and entertainment, where youth culture has long ruled.

At least, it was this way until only recently.

On April 11, 2009, a plain middle-aged woman strutted onto a stage in a London television studio. She gripped the microphone and a rather cocksure man on the adjudicating panel invited her to introduce herself. As she spoke, the man rolled his eyes, barely containing his mirth at the plain woman's brazenness. For this was not just any talent show; it was Britain's Got Talent, the number-one program of its kind in the country. Titters rippled through the hall as the incredulous audience waited for what was surely going to be a cringingly gauche performance.

Then the woman sang.

It took only a few seconds—just the opening lyrics of her song—before the audience fell drop-jawed into a state of shock and disbelief. The woman's voice was so beautiful, so virtuous and so downright surprising that it was difficult to comprehend. Her song, "I Dreamed a Dream" from Les Miserables, rang out with a majestic joyfulness. By the end of her rendition, everyone who'd witnessed it had fallen under the spell of this forty-seven-year-old self-confessed virgin.

Within a few weeks, millions of people had seen Susan Boyle's performance on television, and tens of millions more had watched her on YouTube. She was front-page news for weeks on all major media. By month's end, she was a household name. By the end of the year, Boyle's debut album, *I Dreamed a Dream*, had topped the charts in the US, the UK, Australia and New Zealand, and became one of the fastest selling albums in history. She was now a global celebrity.[87]

Although Susan Boyle is not the first person to experience a meteoric rise to stardom, her case is one of the most instructive, for it exemplifies the nature of the world in which we are now immersed. Firstly, like many other people we have met in this book, Boyle is non-neurotypical—she was diagnosed in late 2012 with Asperger's syndrome, a high functioning form of autism. Secondly, she benefited from an entirely new environment that enabled her dormant "orchid" traits to blossom. Consider for a moment what would have happened if Susan Boyle had delivered her remarkable performance a decade earlier. Certainly she would have surprised many people, but without the viral-like support of YouTube and a host of social networking sites like Facebook and Twitter, it is unlikely that she would have been catapulted to such stratospheric levels so quickly—if at all. To do so then would have taken a lot of hard work by record companies and their publicists to build and sustain her profile. It would also have cost a lot of money, which they might not have been willing to invest in promoting a middle-aged "unknown." In fact, Susan Boyle did perform in public ten years earlier, when she recorded "Cry Me a River" for a charity. Some critics say that the quality of her singing then was as good as, if not better, than her performance on *Britain's Got Talent*. But only 1000 copies of the record were made and Susan Boyle was not heard of again beyond her local village for another decade.

It would take an entirely new media environment to emerge before late-bloomer Susan Boyle could find her place in the limelight—an environment in which a dormant talent, combined with the element of surprise and novelty, could be discovered, celebrated and canonized to an extent that was inconceivable just a few years ago. According to Visible Measures, an American company that computes viewership of Internet videos, Susan Boyle's YouTube video had been watched 410 million times in all of its forms by late 2011, and over half a billion times by 2012. "What we're really seeing with Susan Boyle in a very powerful way is the power of 'spreadability,'" said Henry Jenkins, codirector of MIT's Comparative Media Studies program and author of the book *Convergence Culture: Where Old and New Media Collide*. "Consumers in their own online communities are making conscious choices to spread Susan Boyle around online."[88]

This "spreadability" is facilitated by the unprecedented developments in digital technology and multimedia convergence. We have moved well beyond Andy Warhol's prediction that in the future "everyone will be famous for fifteen minutes." On the web, everyone can be famous to fifteen people—or to 500,000 people—through sites like Facebook, Twitter and YouTube. Those who crave fame—even late bloomers—no longer have to rely on the ephemeral nature of the old media to put them in the spotlight and keep them there. They can make their own spotlight.

## Everyday bloomers

Not every late bloomer craves the spotlight, of course. Many just want to make the most of their newly awakened talent, or simply to live a more fulfilling life. Technology can help them do this, provided they are willing to learn how to use it. But for some elderly people this is easier said than done.

In *Is This Thing On? A Computer Handbook for Late Bloomers, Technophobes and the Kicking and Screaming*, author Abby Stokes described how difficult it was for her own mother to deal with anything technological.

"My mother still can't reset her car's clock after daylight saving time. . . . And the first week after she buys a new car, she only drives it in the Stop & Shop parking lot. Once she feels comfortable enough to take it on the road, it's still a few months before the windshield wipers stop being activated whenever she means to signal a right turn. . . . Mom had never shown any interest in computers, but like so many seniors, she knew she was missing out on something when she began to notice that every article she read ended with 'For more information go to www.[insertanythinghere].com.'"[89]

Stokes, who has helped to demystify computers for over 190,000 people through her book and training courses, suggested to her mother that she think of the computer as a combination of a television set and typewriter, and the Internet as a big library for finding information.

The breakthrough came when her mother wanted to see a Broadway musical. "This was the perfect opportunity to show her what a computer can do," Stokes recalled.

"I turned on my computer, connected to the Internet, and then typed in www.playbill.com [a company that sells theater tickets], and like magic, their website appeared on my screen. I picked the show we wanted to see and the date that was best for us. Next, the seating chart appeared on the screen and we chose our seats. Then I ordered the tickets and typed in my e-mail address where I would receive the e-tickets and then printed them on my printer at home."

Stokes's mother was impressed, and from that day on became an avid computer user and something of a geek. She has used it to find long-lost friends, discover all types of information, book tickets, and generally engage with the world—even when she is housebound. "I am incredibly proud of her for joining the community of computer users," wrote Stokes.

These activities may seem rather mundane to many people, but the point is that Stokes's mother is now engaged with the world in a way that would not have been possible without technology. In her own way, she can bloom again as an active member of her family. "Most of my students are over fifty and are new to computers," Stokes said. "I can't explain to you how exciting it is for them (and me) when they start to zoom around the Internet and begin to enjoy what the digital realm has to offer."

In many ways, the late bloomer personifies the benefits of living in the Anthropocene—a world in which physical limits are transcended through technology. Even though they may not be as mobile and active as before, older people can now access many opportunities to interact, create and engage with the world. The most housebound person can now reach out into the community, participate and have a voice—even if it is just to post a comment online. For those bold souls who wish to start a new career, or make a bigger impact on the world, they too are now more empowered to do so.

## *The app maker* (Omnius ingenius)

Each year, an eclectic group of linguists, historians, editors, writers, and academics meets together under the banner of the American Dialectic Society to choose the "word of the year." It is a serious business for the 125-year-old organization. The judges have to choose from hundreds of contenders for the title. The word does not have to be brand new, but it must have become newly prominent or notable in the past year, in the manner of *Time*'s person of the year.

On January 7, 2011, at a ceremony at the Wyndham Grand Hotel in Pittsburgh, the Society delivered its verdict. The word of the year was "app" (as in software application). In explaining the decision, Ben Zimmer, chair of the New Words Committee of the Society, said, "One of the most convincing arguments from the voting floor was from a woman who said that even her grandmother had heard of it."[90]

Apps are basically tools for getting things done, and range from the practical to the inane. There are apps for calculating, navigating, entertaining, communicating, diagnosing, list making and so on—and they can transform a mobile phone or tablet computer into a digital version of a Swiss army knife, capable of performing a multitude of tasks. Other apps are purely for fun, such as those that invite you to throw angry birds at green pigs, or make fart noises, or invite you to lick the screen. The most pointless app—at least in my opinion—is Hold On!, which asks you to press a virtual hold button for as long as possible, while it counts the time. And that's it.

But make no mistake, apps are a serious business. By mid-2013, fifty billion apps had been downloaded from Apple's app store alone, and billions more were downloaded from the sites of competitors such as Google's Android. Although most sell for a few dollars, the sheer scale and turnover of the industry have made some app developers extremely wealthy.

In April 2012, two young men—Kevin Systrom and Mike Krieger—sold their app called Instagram to Facebook for $1 billion in cash and stock. This simple—yet highly addictive—app enables users to take a photo, modify it, and share it with others on a social network. Over eighty million people now use it. Like many of the geeks we have met in this book, Systrom and Krieger had been students together at a university—in this case, Stanford in California—where they studied computer and engineering subjects. After university, Systrom worked as an intern at Odeo, the company that created Twitter, and then spent two years at Google working on Gmail. Krieger, meanwhile, pursued his interest in symbolic systems, combining software coding with linguistics, psychology and philosophy.

The two men got together again and struck upon an idea that would combine Systrom's love of photography, and Krieger's fascination with symbolic systems, resulting in the new photo-sharing app Instagram, a synthesis of technology and visual art.

In an interview with Somini Sengupta of the *New York Times*, Systrom recalled the moment he and Krieger knew they had a hit on their hands. They were working past midnight in a dimly lit warehouse, known as the Dogpatch Labs, situated on an old pier in San Francisco Bay. Here, amid the fishing-net-draped walls, and surrounded by empty cans of Red Bull energy drink, they made their fledgling app available online for the first time. Within hours, thousands of people had downloaded it, causing the computer system handling the photos to keep crashing. To help them cope with the escalating demand, Systrom called a contact he had met years earlier at a fraternity event at Stanford. His name was Adam D'Angelo, a former chief technology officer at Facebook. He instructed the two budding entrepreneurs on how to keep their system up and running during this critical launch phase. Within the first twenty-four hours, Instagram attracted 24,000 users; by week three there were over 300,000, and within a few months, there were millions of users. Two years later, Facebook came knocking and Systrom and Krieger became wealthy beyond their dreams.[91]

The apps business has also made many other people wealthy too, albeit not all to the same extent as Systrom and Krieger. Beyond the money, however, what's interesting about the "apps gold rush" is that it signifies a historical shift in a long-established industrial model. For the first time in many centuries, individuals are again making "tools"—which is what "apps" really are. Not since before the Industrial Revolution, which ushered in the era of mass manufacturing (requiring molds, dies, jigs, fixtures and so on), have people been able to craft their own tools. For the most part, tool-making has long been an expensive, highly specialized process requiring considerable resources and physical equipment. As we move further into the Anthropocene, however, our tools have become less physical and more virtual. The most important tools we use today are software-based—the programs that run practically every aspect of our modern world. The skills required to produce this software are very different from those of a traditional toolmaker—they are more cerebrally focused, and less physical.

## Myths and mantras

Which brings us to the app geeks. There is a widespread notion—often promulgated by the media—that anyone can design an app, and technically, this is true. But whether it succeeds or not is another matter. App design—like any programming—requires a specific set of cognitive, aesthetic and technical skills. Certainly there have been successful apps created on the cheap, over a kitchen table, by someone with little or no technical skills. But these are the exception. The reality is that for every smash hit like Instagram, there are tens of thousands of failures. A survey conducted in 2012 by the marketing firm App Promo suggested that nearly 60 percent of apps don't break even, and that 80 percent of developers can't make a living from their apps alone. Other research suggests the failure rate is much higher. App developer Paul Kafasis, who works for US software company Rogue Amoeba, said a big part of the problem is visibility. There are over 600,000 apps in Apple's app store, which means that if an app doesn't make the Top 100 or 200 lists, it is unlikely to be noticed. "The App Store is very much like a lottery," he told Chris Foresman at the technology news site Ars Technica, "and very few companies are topping the charts. It's a hit-based business. Much like music or book sales, there are a few huge winners, a handful of minor successes, and a whole lot of failures."[92]

Another challenge, Kafasis said, is that development costs, especially for the more complex apps "are generally much higher than folks realize. Making an app still requires tens of thousands of dollars in development, if not hundreds of thousands. Recouping that kind of money 99 cents—or really, 70 cents—at a time is not easy." Even if an app is well funded, there is no guarantee of success—many investors have lost fortunes on the "next big thing." There is also the added cost of marketing an app after it is launched, which erodes the developer's profit margin.

Notwithstanding the formidable challenges involved, some people do exceptionally well in the apps business. These people usually have a lot in common—they are highly focused, technically orientated and are often part of a close knit fraternity. They are the "app geeks."

A typical example is Phil Larsen, the creator of Fruit Ninja, a game which enables players to slice flying fruit by swiping their fingers across the touch pad, for which they earn points. It is such an addictive game that by mid-2012 it was installed on nearly one-third of all US smartphones, and had been downloaded over three hundred million times, making twenty-six-year-old Larson a multimillionaire. *Time* named Fruit Ninja one of the fifty best iPhone apps of the year. But Larsen is certainly no overnight success or "newbie" (digital newcomer). He worked for many years as a games writer, founded Halfbrick Studios, one of Australia's preeminent video games developers, and has a marketing degree. He even cut his teeth on the archaic Atari 2600—one of the very first video game consoles.

Another success story is Nick D'Aloisio, the British-Australian teenager who sold his Summly app—which summarizes news items—to Yahoo for $30 million in 2013 when he was just seventeen years old. When Summly first caught the attention of the Hong Kong billionaire Li Ka-shing's investment firm two years earlier, D'Aloisio was still in school. At the end of an introductory conference call, the investors invited him to Hong Kong for a follow-up chat and were shocked to hear him respond that it would have to be "out of school hours." Shortly afterwards, D'Aloisio became the youngest person in the world to raise venture capital. Despite his youth, D'Aloisio was not a "newbie" either—he had received his first computer when he was nine years old, began writing apps at twelve years old, and was intellectually gifted enough to win an academic scholarship to the prestigious Kings College School in Wimbleton, UK. He is a natural-born geek.

In a similar vein, there is the example of Markus Persson, the Swedish-born creator of the phenomenally popular game Minecraft—which is also available as an app. Persson started programming on his father's Commodore 128 home computer when he was seven years old, and produced his first game when he was eight. He is also a member of the Swedish chapter of Mensa, the society open only to people with an IQ in the ninety-eighth percentile.

What makes a successful app geek, however, is not just intelligence. We learned earlier in the book that a high IQ is not enough. What is really needed in a global, highly competitive business like app-making is perseverance. For example, Rovio, the makers of Angry Birds, released fifty-two relatively unsuccessful games before they created their smash hit. They just didn't give up. "The companies I talked to all failed massively," explained Chris Stevens, author of *Appillionaires: Secrets from Developers Who Struck It Rich on the App Store*, "they were extremely unsuccessful before they hit on the right app. But here's a different perspective on that: in the world of science, scientists trying to create a successful compound fail hundreds of hundreds of times. So if Rovio failed fifty-two times, it's more of an admirable trait. They persevered."

Perseverance is a trait that geeks do particularly well. They tend to be so obsessively focused that they keep pushing—long after others give up—until they break through. This is especially so when a group of like-minded geeks comes together for a common cause.

Which brings us to another aspect of successful app making—the best developers are usually highly connected to the movers and shakers of the digital ecosphere. They are part of a larger tribe. For example, when Systrom and Krieger set about creating Instagram they were able to tap into a network of investors which included the partners of Benchmark Capital, whom Systrom had met while in college. Through a colleague, they were also introduced to Marc Andreessen, a venture capitalist, who promptly wrote Systrom a check for $250,000 to help kick-start the new venture. When it came to expanding their business, Systrom and Krieger were able to employ highly skilled geeks from their own circle of acquaintances.

Systrom acknowledged the importance of networking, when he addressed a group of would-be entrepreneurs at his alma mater in 2011. "Make sure to spend some time after [this talk] getting to know the people around you," he urged the students, the implication being that one day they might need these contacts—as he once did in reconnecting with D'Angelo.

For all their individual brilliance, geeks often work best in a group situation. They need to be able to bounce ideas off like-minded people, as we saw earlier in this book. They thrive in a hothouse environment where their special orchid-like traits can bloom. This was certainly the case for the creators of Instagram, who found themselves in the right place at the right time. Indeed, much of the success of Systrom and Krieger, according to Somini Sengupta, has to do with "the culture of the [San Francisco] Bay Area tech scene, driven by a tightly woven web of entrepreneurs and investors who nurture one another's projects with money, advice and introductions to the right people. By and large, it is a network of young men, many who attended Stanford [University] and had the attention of the world's biggest venture capitalists before they even left campus."

Through a combination of ingenuity, perseverance and networking these geeks have helped to reinvent the way we interact with the world. They have created apps—modern-day tools—which are transforming how we communicate, navigate, plan, record and share our experiences. They provide us with a sense of empowerment by making us feel that things are more manageable and predictable. I recall waiting for a taxi in London during chaotic peak-hour traffic in pouring rain. Normally I would have been frustrated, and concerned that the taxi would never show up. But I had downloaded a marvelous little app that indicated the location of my taxi as a dot on the screen—and I could see it was creeping inexorably toward me. The app didn't actually change the situation, or shorten my waiting time, but it made me feel that things were under control. A lot of apps are like that—even the fruit-slicing, bird-throwing variety. They enhance our sense of personal mastery over our environment.

Perhaps most importantly, however, these apps can make us feel more intelligent. They extend our cognitive and perceptual abilities, by enabling us to perform mind-based tasks much faster than we could do unaided. Google's "chief technology advocate," Michael Jones, told the *Atlantic* in 2013 that "effectively, people are about 20 IQ points smarter now because of Google Search and Maps. They don't give Google credit for it, which is fine; they think they're smarter, because they can rely on these tools."[93] Jones's claim may be an exaggeration, but there is no doubt that apps can make us feel more intelligent.

This brings us to the next chapter of this book—"The Cognitive Revolution"—in which we shall explore how new technologies are about to transform our ability to think.

# Chapter Four

## *The Cognitive Revolution*

So far in *Unnatural Selection*, we have seen how the Anthropocene has encouraged the rise of the geek by creating a "digital greenhouse" that favors their unique cognitive traits. This gives them an advantage in many techno-centric activities, and has enabled some to achieve phenomenal success. We have analyzed various types of geeks, contemplated their motivations, and considered the consequences of their behavior. We have seen how their attitudes to issues such as privacy and the treatment of information (facts) affect the way we interact, particularly in social media. The standards they set—technically and behaviorally—help shape the protocols of the modern world.

In many ways the geeks are a personification of the Anthropocene. As such, they act as powerful role models and influence the values that underpin our society. There is one value in particular that they have elevated above all else—intelligence. The geeks are nudging us—directly and indirectly—to become a more intelligent society. One that celebrates cognitive ability.

This is not to suggest, of course, that we no longer enjoy less cerebral physical activities such as sport, shopping, travelling and sex. Rather that, increasingly, we do everything with an intellectual—or cognitive—bias.

So, for example, many of us don't just go jogging anymore; we simultaneously check our progress on a wearable cardio monitor that tracks our speed and position via GPS. Instead of wandering through department stores, we shop online with price comparison devices. Even sex, the most physically intimate of all pastimes, is becoming more virtualized through Internet porn. Whatever we do in the physical world, we now increasingly filter our experience through a computer screen, which in turn engages our cognitive functions.

In effect, we are becoming more geek-like in our day-to-day behavior. We have no choice, really, because the only way to keep up with the world is to become more technologically proficient. This applies as much to children as it does to late bloomers. No one wants to be left behind. In fact, many people don't just want to keep up, they want to get ahead. They are eager to join the cognitive elite in order to better navigate the complexities of the Information Age.

But it's one thing to be able to operate an iPad, and quite another to be a genuine geek. As we learned in chapter two, "Unnatural Selection," geeks are a relatively rare breed—orchids in a field of dandelions—and not everyone can be one. It is also unlikely that the super-geek population will increase because, like many intelligent people these days, they tend to have fewer children. So their numbers won't grow significantly—at least not through the evolutionary process of natural selection, which is all about reproductive success. Certain things are about to change, however. Our society is undergoing a cognitive revolution that will enable practically anyone to have geek-like focus and attention. It's a revolution being driven by breakthroughs in neurobiology, pharmacology, genetic engineering, augmented reality and artificial intelligence. These cognitive enhancements won't just extend our intelligence—like, say, an app—but will fundamentally transform it by enabling us to think faster and more clearly and be more creative. They offer a shortcut to geekdom and will accelerate our evolution into a more cerebral society.

If this sounds far-fetched consider that it was barely a few decades ago, in 1978, that Louise Brown, the world's first "test tube" baby, was born. By the time she turned eleven years old, scientists had moved beyond creating life in a laboratory and were capable of looking directly into an embryo's genes in order to predict what the future might hold for that life. This meant that parents could preview, and select, some of the characteristics of their intended children. By 2002, the entire human gene was decoded, which has subsequently triggered an explosion in genetic and stem-cell research. Now we are on the brink of an epigenetics revolution, which will enable us to deliberately influence the way our genes are expressed. All this astonishing progress has occurred in the blink of an eye—in evolutionary terms—and is accelerating.

Of course, humans have been interfering with evolution— including our own—for a long time. We have bred crops and livestock for food, which in turn has modified our ability to handle carbohydrates and proteins. We have bred horses for transport, dogs for hunting, sheep for wool, and pigeons for communication, to name just a few examples of our tampering with nature. In our darkest hours we have even tried to breed humans with certain physical characteristics, through ill-fated eugenics programs, such as those practiced by the Nazis during the 1930s and '40s, when they attempted to produce a race of blue-eyed Aryans. But, for the most part, our evolution has occurred according to the laws of natural selection. It has not been guided in any particular direction; it has operated in a mindless and mechanical way. It just happened. This is the process Richard Dawkins has described as analogous to a "blind watchmaker."

Now we are at an evolutionary crossroads. We are increasingly interfering with our own evolution—whether directly through assisted reproduction techniques (such as IVF) and medical interventions, or indirectly by altering our environment. We cannot pretend to ourselves that the "watchmaker" is completely blind anymore. The first thing we are likely to do with this power to direct our own evolution is improve our intelligence. For this is our most valuable human asset, and distinguishes us from every other life form on the planet. It has lifted our species to great heights. Cognitive enhancement technologies could lift us even higher and faster, and offer many benefits to society—socially, culturally and economically. Studies published by the Oxford Centre for Neuroethics estimate that a 3 percent population-wide increase in IQ would reduce poverty rates by 25 percent, leading to an annual economic gain of between US$165 and $195 billion and a 1.2 to 1.5 percent increase in GDP.[1] Access to cognitive enhancement could help developing countries to participate more fully in the global economy. It might also reduce inequality and promote social justice.

The most powerful impetus for the cognitive revolution, however, is likely to be good old-fashioned self-interest. As discussed earlier, people want to be smarter in order to succeed in a highly competitive world. To join the cognitive elite they might be willing to take evolutionary shortcuts, and to pay a high price.

Let's now explore this cognitive revolution—or pursuit of geekdom—in some detail. As you will see, we humans are about to embark on a remarkable journey of self-transformation that is both exhilarating and somewhat scary. What you might find surprising—as I certainly did—is how far we have already come along this path.

## Plastic brains

Until quite recently, it was assumed that by the time we reached adulthood, our brain anatomy would no longer continue to develop but would remain fixed for the rest of our lives. This static view of the brain gave rise to a "neurofatalism" that severely limited our view of how much control we had over our mental faculties. It was like being told, "Here is your 1990 computer. Get used to it because it is the only one you will ever have . . . and, by the way, it can't be upgraded."

This fixed perspective of the brain also had wider implications, for it suggested that human nature itself, which arises from the brain, is inflexible and unalterable, and therefore not worth trying to change. We are slaves to our neurological destiny.

This view was driven in large part by the idea that the brain was really a machine, and therefore did not continue to evolve after it was created. It was also a very fragile and complex machine, so that if a part broke down it could not be fixed. This perspective appeared to be supported by the observation that brain-damaged patients rarely fully recuperated.

This view of the brain began to fracture during the 1990s when scientists and doctors began reporting remarkable stories of patients who had overcome serious neurological disorders. Simultaneously, advances in brain imaging technology, such as

MRI scans, were able to show distinct changes in brain anatomy over time. It became clear that the brain is not fixed, after all. In fact, it is the opposite—it is highly malleable or "plastic." Every time we use our brain it makes new connections between the cells and creates new cells (neurons) through a process of neurogenesis. Our brains are constantly rewiring themselves.

This remarkable insight began to percolate through the medical fraternity, where it bubbled away for a number of years, seeping into peer-reviewed papers and conferences. Then in 2007, a couple of ground-breaking books catapulted the issue into public consciousness. In particular, *The Brain That Changes Itself* by psychiatrist and psychoanalyst Dr. Norman Doidge made it clear just how malleable the brain is.[2] He recounted in the book how, during the course of his research, he met scientists who had enabled blind people to see again, and deaf people to hear—even though they had been afflicted since birth. He went on:

"I spoke with people who had had strokes decades before and had been declared incurable, who were helped to recover with neuroplastic treatments; I met people whose learning disorders were cured and whose IQs were raised; I saw evidence that it is possible for eighty-year-olds to sharpen their memories to function the way they did when they were fifty-five. I saw people rewire their brains with their thoughts, to cure previously incurable obsessions and traumas. I spoke with Nobel laureates who were hotly debating how we must rethink our model of the brain now that we know it is ever changing."

In the same month that Doidge's book came out, *Train Your Mind, Change Your Brain* by science writer Sharon Begley appeared, echoing the theme of neuroplasticity and highlighting the beneficial effects of mental exercises—particularly meditation—on brain function.[3] It showed how regular Buddhist meditators consistently achieved higher baseline measurements for positive emotions, such as compassion and general well-being. Other recent studies conducted by researchers at Yale, Harvard and MIT have provided evidence that the regular practice of "mindfulness" meditation (which emphasizes focusing on the present moment) increases the thickness in those parts of the brain that deal with attention and processing input. Normally, these areas within the human cortex (or thinking cap) would get thinner as we age. Sara Lazar, the leader of one study and a psychologist at Harvard Medical School, told the Harvard Gazette:

"Our data suggests that meditation practice can promote cortical plasticity in adults in areas important for cognitive and emotional processing and well-being. These findings are consistent with other studies that demonstrated increased thickness of music areas in the brains of musicians, and visual and motor areas in the brains of jugglers. In other words, the structure of an adult brain can change in response to repeated practice."[4]

Importantly, the meditators involved in these studies weren't monks living in seclusion who had devoted decades of their lives to contemplation. Most of them worked in busy careers such as law, health care and journalism, and meditated for no more than forty minutes a day.

As Dr. Jon Kabat-Zinn, one of the pioneers in bringing mindfulness meditation into mainstream medicine, and founder of a mindfulness-based stress reduction program, has remarked, Eastern traditions have long been interested in developing the human mind. In his book *Wherever You Go, There You Are* he noted that meditation is often referred to in Buddhist scriptures as *bhavana*, which translates as "development through mental training." To me, this strikes the mark; meditation really is about human development.[5]

The discovery of neuroplasticity has prompted entrepreneurial educators to create new programs designed to enhance cognitive functioning. Many of these can be accessed online, for example, the popular brain training program Lumosity, which provides game-based exercises that target different mental processes, such as memory, focus and attention. Another product is Mindsparke, which aims to increase mental flexibility, IQ and creativity. Like meditation, however, the success rate of these programs is highly dependent on the effort made. The more you put in, the more you get out.

Many people, however, are not willing to make the effort. They are looking for a shortcut.

## Smart pills

Humans have always taken drugs to alter their states of mind. When people first nibbled plants, they noticed that some would exert a curious effect over their senses. They soon learned that these exotic plants could be used to lift their mood, enhance enjoyment and reduce pain. They could even affect consciousness, in ways that proved useful for tribal ceremonies and spiritual practices. These early drugs, humans discovered, could help them to adapt to the physical, cultural and spiritual needs of their time.

It is no different today.

The demands of the Anthropocene, with its strong emphasis on conceptual and analytical skills, are encouraging people to find new ways to enhance their cognitive functioning. Many are turning to "smart pills"—or neuro-enhancers—to improve their brain power. Among the most commonly used smart pills today are Adderall (mixed amphetamine salts), Ritalin (methylphenidate), Vyvanse and Provigil (modafinil), which can boost attention and memory capabilities. According to a recent study of nearly 11,000 university students in the United States, about 7 percent of them reported having taken such stimulants "non-medically"—mainly to improve their grades.[6] Another study by Barbara Sahakian, professor of clinical neuropsychology at Cambridge University, found that "17 percent of students in some US universities admitted using the stimulant Ritalin—a drug designed to treat hyperactive children—to maximise their learning power."[7] Students believe that if these neuro-enhancer drugs can improve their memory function by just a small margin, it could be the difference between passing and failing an exam, between a good grade and a better one. Professor Sahakian says that students can feel pressured to use these drugs to keep up with their peers. Unlike the usual profile of drug takers, these kids are usually A students, and sometimes B students, who are trying to obtain good marks.

Most of these drugs are not manufactured to be cognitive enhancers—they are being used "off label"—that is, for a different purpose from what is described on the medicine's packaging. Ritalin, for example, was created to calm people with ADHD, but it can have the opposite effect on people without the disorder. A single pill can turbocharge these users' energy levels and enable them to push through all-night study sessions, or arduous exams. "It's like [the drug] does your work for you," a recent graduate of the Birch Wathen Lenox School on the Upper East Side of Manhattan told the *New York Times*.[8]

It is perhaps little wonder that the number of prescriptions issued in the United States to young people aged ten to nineteen for medications such as Adderall has risen by 26 percent since 2007, to almost twenty-one million yearly, according to the health care information company IMS Health. While most of these prescriptions will be for legitimate reasons, an increasing proportion of them are being used for cognitive enhancement.

It's not just students using them either. Dr. Anders Sandberg, a philosophy lecturer specializing in bioethics at Oxford University, told the *Daily Mail* newspaper that he quite openly uses modafinil as a cognitive enhancer.[9] "When I take it, it is like having a little electric motor in the back of my head running through lists of things I need to do," he said. "Then, instead of putting them off until tomorrow, I go ahead and do them. I use the drugs only occasionally if I have a paper to write or need to fly long distances to attend a conference or deliver a speech. I find that instead of having jet-lag, I can focus on the job at hand."

As a bioethics philosopher, Sandberg has grappled with the message he is sending to students. "This is something I have spent a lot of time considering, but in general I believe people should have control over their own bodies," he said, although with a caveat. "These drugs are like step-ladders. If you need them to attain something that would otherwise be out of your reach, then use them. But if you can reach those heights anyway, then you're just being lazy." It could be argued, however, that for some students "being lazy" is precisely the problem that the drugs enable them to overcome.

Beyond the campuses, the cognitive revolution is infiltrating other walks of life and professions. Truck drivers, nurses, doctors, shift workers and pilots are taking stimulants to beat fatigue and enhance alertness. Academics, writers and creative people take drugs like modafinil to keep them awake long enough to meet their deadlines. Military personnel use them to stay sharp—particularly Special Forces operatives who can be required to undertake arduous missions lasting over forty-eight hours without sleep.

Business executives too, who have long used beta-blockers to calm their nerves, are now taking cognitive enhancers to boost their performance. A typical example was described by Maia Szalavitz, a US neuroscience reporter (who also cited the study of 11,000 US students), of forty-year-old "Bob," a high-level e-commerce executive, who felt he was losing his edge.

Although his colleagues saw him as a star, he feared he wouldn't be able to continue the lightning pace and constant multitasking his job required. So he saw his doctor. Now Bob takes Adderall, a prescription amphetamine ordinarily used to treat attention-deficit/hyperactivity disorder (ADHD). "[The drug] gives me clarity of thinking and focus," says Bob. He credits the drug with improving both his career and his personal relationships. "I am still getting accolades," he says.[10]

What drugs like Adderall, Ritalin and modafinil do, in the words of Professor Sahakian, is to bring "improvements in complex planning and problem-solving tasks, namely the executive functions in the front part of the brain. Modafinil has also been shown to improve memory functions and Ritalin has been shown to specifically improve working memory."

These smart drugs are known as nootropics (a term coined in 1972 by a Romanian doctor, Corneliu E. Giurgea), from the Greek *nous*, or "mind," and *trepein*, meaning "to bend" (that is, mind-bending). They are believed to work by modifying the availability of the brain's supply of neurochemicals, and by stimulating nerve growth and improving oxygen supply to the brain.

The pharmaceutical industry is now investing hundreds of millions of dollars in nootropics—or smart pills—in the hope of creating products that not only relieve neurological disorders, but also enhance cognitive functioning. Some of this investment is from governments seeking to exploit advances in neuroscience technology for applications in public health, the military and national security.

This will lead to entirely new generations of drugs designed specifically to make people smarter. They will no longer be used "off-label," but actually be prescribed for the purpose. Such drugs will be easy to market in an age when many people are struggling to cope with the intellectual demands of the Anthropocene.

A smart pill will offer them a shortcut to "geekdom."

A growing number of scientists and psychologists, however, while recognizing the benefits of neuro-enhancing pills, are concerned about their potential side effects. These effects are not just physical and mental, they are also sociological, and have political repercussions. Many people regard cognitive enhancement as unnatural, a rationalization of drug abuse, or just plain cheating. They argue that because doping is banned in competitive sports such as the Olympics, it should also be banned in college and university exams—otherwise the contest is unfair. Students, like athletes, should be "clean" to compete.

There are also the behavioral implications. Tom Newham, a student writing in the *Guardian*, described how modafinil affected some of his friends in the lead-up to exams:

"Over those weeks [while they were taking modafinil] my friends became different people—in turn aggressive, cold and reclusive. Eating was 'a waste of time' and so was conversation. One friend, a world-class procrastinator, could be found swearing at anybody who interrupted his workflow, walking away from conversations mid-sentence. When I put it to another that using brain-enhancing drugs amounted to cheating, he turned on me, accusing me of wanting to ban revision. He apologized the next day. He said it was the drugs talking."[11]

Smart drugs may be particularly problematic for children and teenagers. Paul L. Hokemeyer, a family therapist at Caron Treatment Centers in Manhattan, told the *New York Times*, "Children have prefrontal cortexes that are not fully developed, and we're changing the chemistry of the brain. That's what these drugs do. It's one thing if you have a real deficiency—the medicine is really important to those people—but not if your deficiency is not getting into Brown [an elite US college]." There is also the problem of legality. Most students don't realize that their so-called study drugs are in fact illegal to use without a doctor's prescription, which could land them with a criminal record. They are not like vitamins.

Meanwhile, long-suffering parents are now faced with yet another drug problem. Dodi Slar, the mother of a child attending Ardsley High School in New York (which Mark Zuckerberg once attended), listened to her ninth-grade son describe how some classmates were already using stimulants to improve their grades, while assuring her that he would not do it. She told the *New York Times*, in the same article, "You worry about [their] driving, you worry about drinking, you worry about all kinds of health and mental issues, social issues. Now I have to worry about this, too? Really? This shouldn't be what they need to do to get where they want to."[12]

There is also the problem of safety. Many people assume that because a drug is available in a pharmacy, it must be OK for anyone to take—even if it is used "off label." This is not true. Adverse reactions to commonly prescribed drugs are a leading cause of death, disease and disability. In the United States alone they account for over two million hospitalizations and 110,000 hospital-based deaths a year, not including those occurring at home or in nursing centers.[13]

The point is that not everyone will respond to cognitive enhancers in the same way, no matter how well designed they are. Some people will experience adverse reactions. According to Rob Brooks, who is professor of evolutionary biology at the University of New South Wales, "Some individuals will thrive on these new smart drugs, and become 'super geeks,' and others will not. The naturally occurring genetic differences in people will mean that some people benefit hugely, while others hardly benefit at all. Some might even suffer—maybe they will get seizures and some will die [just like some ecstasy users do]."[14]

There is also the issue of fairness. Smart drugs are not cheap—and it is likely they will become more expensive as pharmaceutical companies invest more money in their development—money they will want to recoup. Also, the fact that many people would pay almost anything to get an edge over their competitors, this may drive up prices, so that only the wealthiest—or best connected—people will be able to afford the latest smart pill. This would perpetuate the development of a cognitive elite, which, over the long term, would further exacerbate the gap between rich and poor. It would undermine the societal advantages of increased intelligence discussed earlier, because relatively few people would benefit.

Notwithstanding these challenges, it is inevitable that smart drugs will become more commonplace, and acceptable. In 2009, a group of seven leading bioethicists and neuroscientists published an editorial in *Nature* advocating the use of performance-boosting drugs, which stated: "Society must respond to the growing demand for cognitive enhancement. That response must start by rejecting the idea that 'enhancement' is a dirty word."[15] The authors made the case that even though smart drugs alter brain function, so do many other enhancing activities such as nutrition, exercise and sleep, and that therefore, "cognitive-enhancing drugs seem morally equivalent to other, more familiar, enhancements."

They suggested that given the potential benefits of cognitive enhancement to individuals and society, "a proper societal response will involve making enhancements available while managing their risks."

The *Nature* authors did, however, offer some caution, suggesting, it would "be foolish to ignore problems that such use of drugs could create or exacerbate." With this, as with other technologies, we need to think and work hard to maximize its benefits and minimize its harms. One of the paper's authors, Martha Farah, director of the Center for Cognitive Neuroscience at the University of Pennsylvania, told *Time*, "If it were possible to call for a moratorium on cognitive enhancement until the risks are better understood, that would obviously be the best thing to do," but "the genie is already out of the bottle."[16]

Smart drugs are here to stay and will exert a powerful influence on our cognitive capabilities. They may even influence human evolution itself, as Rob Brooks explained:

"The genes that—over time—work best in the new smart-drug saturated environment of the Anthropocene will eventually increase in frequency and those genes that won't mesh well with the drugs will be lost. This may eventually have an impact on the way we evolve. So we—as a society—may become more geek-like."[17]

Cognitive enhancement, however, is not limited to drugs.

## Mind your genes

Ever since the Human Genome Project was completed in 2003—identifying 20,000 to 25,000 genes in the human DNA—the race has been on to decode what each of these genes does. Much of the focus has been on identifying which ones cause disease and how they can be modified to avoid it.

The cofounder of Google, Sergey Brin, has taken a particular interest in this research. When he turned thirty-five years old, he learned that he had inherited from his mother a genetic mutation called LRRK2, which appears to predispose carriers to Parkinson's disease. Brin told *The Economist* that he regards his mutation of LRRK2 as "a bug in his personal code," and as such it is no different from the bugs in computer code that Google's engineers fix every day.[18] He hopes that one day everyone will be able to learn their genetic code in order to be able to take proactive measures to ward off diseases, or try to repair their genetic bugs. A number of companies are already offering genetic-analysis services to the public, including 23andMe, which was cofounded by Brin's wife, Anne Wojcicki, who is a biotech analyst.

It is only a matter of time, however, before genomic scientists look beyond the search for bugs in the personal code that cause disease, and turn their attention to genetic enhancement—that is, the manipulation of human genes to produce superior characteristics.

In the 1997 film *Gattaca*, we are given a preview of a dystopian future in which embryo screening is routinely used to ensure that children are born with the optimum genetic mix. In one scene, a geneticist reassures a prospective parent, "This child is still you, simply the best of you. You could conceive

naturally a thousand times and never get such a result." The
film explores the moral ambiguities of such unnatural selection
through the intersecting lives of two young men—one (played
by Jude Law) born genetically gifted, and the other (played by
Ethan Hawke) who is born naturally without genetic screening.
At one point, the Ethan Hawke character, who is desperate to
become an astronaut but is limited by his genes, says, "I belong
to a new underclass, no longer determined by social status or the
color of your skin. No, we now have discrimination down to a
science. . . . They used to say that a child conceived in love has
a greater chance of happiness. They don't say that anymore."

At the time *Gattaca* was released it was considered to be
pushing the boundaries of what was technically feasible, even
by Hollywood standards. The significance of the film's title
was lost on most people (except biologists), for it is composed
of the initial letters of the four DNA nitrogenous bases (ade-
nine, cytosine, guanine and thymine), which are the building
blocks of life.

Nowadays, given what we know about genetics, the premise of *Gattaca* is not only realistic, but conservative. As if to
highlight the point, a recent study published in *Nature Chemical Biology* by scientists at the Scripps Research Institute in La
Jolla, California, suggests that it will be possible to introduce
brand-new letters into the genetic alphabet beyond the original
Gattaca list.[19] This would enable scientists to code for a much
wider range of molecules and create powerful applications,
from precise molecular probes and nano-machines, to useful
new life forms.

There is one area where genetic engineering is destined to
play a major role—cognitive enhancement. This is not surprising, given that genes determine at least 50 percent of a person's

intelligence. Scientists are racing to discover which genes affect intelligence, how they operate, and how their function can be improved. Some neuroscientists refer to this emerging field as "plastic surgery for the brain"—or "cosmetic neurology"— where you periodically get a tune-up of your synapses to make you appear sharper, smarter and on the ball.

However, as we have seen, just because a person has the right genes doesn't necessarily mean they will apply their intelligence and become successful. Much depends on environmental influences. But bright genes will certainly give them a head start and a competitive edge in the Anthropocene. They will be the cerebral equivalent of an athlete who is born with a genetic mutation that responds positively to the hormone erythropoietin. Such athletes can carry more oxygen in their blood—thus enabling them to excel at endurance sports. Similarly, people with genetically enhanced minds could expect to have more mental endurance and to be able to focus more clearly, more efficiently, and for longer periods. They will be granted automatic membership of the cognitive elite.

## Neuro implants

The National Spelling Bee of 2023 started out like any other, but controversy enveloped the contest when Suzy Hamilton, an eight-year-old from Tulsa, emerged as the new champion. Contestants had been getting younger for years; that was nothing new. But midway through the event it was discovered that Suzy was—in the words of one commentator—"amped." At the age of four, suffering from seizures and severe attention and behavioral problems, Suzy had received an experimental new treatment: a neural implant that prevented her seizures and helped her to focus. As it turned out, the device also appeared to make her a prodigy at memorization, as her parents and teachers soon discovered.

This depiction of a cognitively enhanced girl is by Daniel Wilson, whose futuristic novel *Amped* explores the consequences of a world in which people's mental faculties are technologically boosted.[20] Wilson, who has a doctorate in robotics from Carnegie Mellon University in Pittsburgh, Pennsylvania, has suggested that we are on the verge of creating a new class of super-enabled people with microchip-enhanced mental abilities. Writing in the *Wall Street Journal*, he proposed that:

"Over the next decade, new implantable technologies will fundamentally alter the social landscape. We are fast approaching a milestone in the eons-long relationship between human beings and their technology. Families once gathered around the radio like it was a warm fireplace. Then boom boxes leapt onto our shoulders. The Sony Walkman climbed into our pockets and sank its black foam tentacles into our ears. The newest tools are creeping still closer: They will soon come inside and make themselves at home under our skin—some already have."[21]

Neural implants are usually no bigger than pill-sized and attach to the brain just under the skull. Some, known as deep brain stimulators (DBS), are already being used to stimulate parts of people's brains stricken by inactivity caused by strokes, head injuries and neurological afflictions such as Parkinson's disease. They act as brain pacemakers. Tens of thousands of Parkinson's sufferers worldwide have undergone a DBS procedure, which is relatively straightforward and usually involves an overnight stay in a hospital. A study conducted by the UK's Institute of

Neurology's Unit of Functional Neurosurgery demonstrated that at their twelve-month follow-up visits, patients' symptoms and signs of Parkinson's had improved by 55 percent compared with their disability at the start of the study.[22]

Other types of neural implants can help compensate for a loss of vision, hearing, or touch. Future devices could record our entire life experience as a continuous video feed, which would serve as a backup if our own memory fails. Scientists are also working on creating "quantum dots," which are microscopic photosensitive flecks of silicone that could be used to inject new information into the brain and provide noninvasive treatment for Alzheimer's, epilepsy and blindness.

It may also be possible for future implants to "hotwire" damaged brain circuits and boost performance. In October 2012, a research team at Wake Forest University School of Medicine in Winston-Salem, North Carolina, announced that they had developed and implemented a neural implant in primates, designed to improve higher-order brain processing.[23] Their implant was able to successfully stimulate and record activity in multiple levels of cells in the prefrontal cortex—the decision-making part—of the primates' brains. Most significantly, they were also able to increase the speed and accuracy of the monkeys' thinking processes.

Future generations of implants will create interfaces between the brain's neural systems and computer chips—thus enabling people to directly access and process vast information databases just by focusing on them. In effect, their cerebral cortex—the most highly evolved part of the brain—will have been upgraded. Implants will also help monitor and manage our ability to concentrate and pay attention.

"These tools aren't sinister," wrote Daniel Wilson. "They're being created to solve real problems. Simply put, prosthetic limbs help people move, and neural implants help people think." He is concerned, however, about the impact and fairness of such cognitive enhancements, as highlighted in his hypothetical example of eight-year-old Suzy Hamilton:

"'Was this youngest-ever winner [of the spelling bee] fully human or was she part machine?' he pondered. 'Was it fair for her to be in a competition with peers made of mere flesh and blood? When your child is the only kid in her class without an implant and she has the lowest test scores to prove it, will you agree to put her under the knife?'"

Nevertheless, Wilson believes that the benefits of technologies such as neural implants outweigh the downside, because they can correct imbalances in society and unleash hidden potential. "The person who has a disability today may have a super ability tomorrow," he suggested.[24]

## Augmented reality

Another technology destined to enhance our cognitive abilities is "augmented reality," which will help us interact with the environment around us by overlaying our view of the real world with digital information.

Augmented reality systems have been around for years in the form of "heads up displays" worn by fighter pilots, which provide them with vital systems data without the need for them to take their eyes off the windshield. The latest version on the USAF F-35 Lightning II even allows the pilot to look through the aircraft's walls at an uninterrupted view of the world out-side. Augmented reality systems have now been introduced into commercial aircraft, and more recently into cars, to assist drivers with navigation, traffic hazards and so on.

Such systems are about to become a lot more personal in the form of wearable glasses, which project information directly in front of the user's eyes. The idea is that the wearer of such glasses will be able to walk along a street and see before their eyes street directions, incoming emails, messages, news and weather updates. Look at a building and the glasses could provide its name and its opening and closing times. Chat with a local in a foreign country and real-time translations could appear before the wearer's eyes — without the other person even knowing. The glasses would help answer difficult questions on the spot, and give the wearer details about a person they meet, by, for exam-ple, consulting the individual's Facebook or LinkedIn pages. In other words, these "geek glasses" could make a person operate more intelligently and more responsively to their environment. They raise their wearer's "situational IQ."

Admittedly, the first generation of augmented reality glasses introduced by Google in 2014 do appear rather cyborg-ish. But this will inevitably change as they become indistin-guishable from normal glasses. The intention is, as Sergey Brin has said, that the wearer — and those around them — will barely notice they have them on. "The best technology," he told the annual Google Developer's Conference, is the type that "gets out of the way." He described playing with his young son Benji by throwing him in the air with both hands and catching him; as he did so, his Google glasses took pictures of what he was seeing in the moment. "I could never have done that with a smartphone or a camera,"[25] The *New York Times*'s Nick Bilton had the opportunity to try the Google glasses and described his impressions in his blog:

> The experience was as mesmerizing as when I saw the iPhone for the first time. The screen of Project Glass [Google glasses] sits off to the side, clear and unobtrusive. You interact with it when you need to. When an e-mail or text message comes in, you can look if you want, or simply ignore it. It's not as if a large red stop sign is jammed in your face when messages arrive. These things obviously have their share of problems. They cost $1,500 for a special pre-order. Although I'd gleefully walk around with a pair on, my sister and the majority of readers of this newspaper, would probably say, "No thanks. Way too geeky for me." But that will all change.[26]

Another person who tried the glasses is Joshua Topolsky, an American technology journalist, who told the *Guardian*, "In the city, Glass [Google glasses] makes you feel more powerful, better equipped, and definitely less diverted." But, he added, "It might not be that great at a dinner party."[27]

As augmented reality glasses take off, it will be interesting to see what effect they have on people's behavior, and whether they affect communication habits in the way that, say, text messaging has encouraged people to be more concise (sometimes, rudely so). It is not difficult to imagine a scenario in which people's conversations become partly scripted by their smart glasses. One can imagine, a shy young man on a first date consulting his glasses for some witty repartee, or a nervous job interviewee relying on her glasses to help respond to questions more coherently.

There is a distinct danger that over time augmented reality glasses may encourage us to become masters of the autocue—like

TV presenters or politicians—which would undermine spontaneity and authenticity. This in turn, might prompt new social etiquette rules, as the introduction of mobile phones did. Cafés and bars might erect signs saying, "Natural Conversations Only, No Augmentation Allowed." Nevertheless, augmented reality is well on its way, and not just from Google. Other manufacturers, including Apple and the Japanese optics manufacturer Olympus, are racing to release their versions of smart glasses. Some are even developing contact lenses with similar capabilities. Industry analysts, such as Forrester Research, have suggested that augmented reality devices will be the next big thing, after smartphones and tablet computers. Such devices will help us to navigate the growing complexities of the Anthropocene, and we may end up relying on our smart glasses as reflexively as we reach for a calculator to do math.

## Artificial intelligence

Of all the cognitive-enhancement technologies on the horizon, the most promising is the development of artificial intelligence (AI), which means computers that "think." AI technology will be used initially to boost the effectiveness of the enhancements we have discussed, such as smart glasses and neural implants. But over time, it is likely that various forms of artificial intelligence will be integrated directly within our own neural networks, thereby transforming more people into genuine geeks—and accelerating our evolution into a truly cerebral society.

Recent advances in the field of AI are, excuse the pun, mind-boggling. They are also somewhat scary. Researchers at IBM and DARPA (the US Department of Defense Advance Research Projects Agency) have created experimental "brain chips," which emulate the working of the human brain by forming, reforming and strengthening artificial synapses — equivalent to neural synapses. So instead of being calculators, which most computers essentially are, these brain chips can, theoretically at least, learn from their experiences and reshape themselves through an artificial version of neuroplasticity.

The significance of this development is that over time these chips may enable artificial intelligence to evolve from the ground up, rather than the top down. Until recently it was assumed that artificial intelligence would have to be designed in the same way that a computer is. That is, we would know what it looked like, how it operated and why it did the things it did, because we had created it from a top-down perspective. But if such intelligence can emerge over time — from the bottom up — through a technological version of natural selection, we would have no idea how it would turn out in the end.

This revolutionary perspective on AI was recently articulated by one of the world's foremost philosophers of science, Daniel Dennett, who suggested that it is likely that artificial intelligence would evolve in a similar way to biological intelligence. Writing in *The Atlantic* in July 2012, he explained that in the pre-Darwinian world we assumed that everything was created by an "omnipotent and omniscient intelligent creator — who bore a striking resemblance to the second-most exalted thing [us]. Call this the trickle-down theory of creation. Darwin replaced it with the bubble-up theory of creation [that is, evolution]."[28] Similarly, Dennett says, the computer pioneer Alan Turing overturned the trickle-down theory of artificial intelligence when he realized that all computational problems could be broken down into a few simple steps — akin to what happens in the natural selection process:

As Turing fully realized, there was nothing to prevent the process of evolution from copying itself on many scales, of mounting discernment and judgment. The recursive step that got the ball rolling—designing a computer that could mimic any other computer—could itself be reiterated, permitting specific computers to enhance their own powers by *redesigning themselves* [emphasis in original], leaving their original designer far behind. . . . . [Turing] recognized that there was no contradiction in the concept of a [non-human] computer that could learn.

Which brings us back to IBM's brain chips. If Dennett is right, then these chips—or future iterations of them—could, over time, evolve to actually think in a way that approximates general artificial intelligence. The more that the chips are used in everyday applications—such as monitoring water quality or grocery food stocks—the more opportunities they have to evolve. Computer evolution—like biological evolution—is all about cycle frequency (measured in generations), the number of connections, algorithms and processing power.

One of the most convincing examples of how AI might evolve in this way was recently demonstrated at Google's secretive X laboratory—which is known for creating self-driving cars and the augmented reality glasses discussed earlier. The laboratory team connected 16,000 computer processors to create what was in effect one of the world's largest artificial neural networks, with over one billion connections. They then exposed this network to about ten million digital images found in YouTube videos. The objective was to see if the "Google brain" could mimic some of the functionality of the human brain's visual cortex, which has evolved to recognize patterns that are meaningful to us (such as faces and their expressions).

Within a relatively short time, the Google brain was able to recognize cats (which are a popular YouTube theme), and also homed in on human faces.

"So what?" you might ask. Surely any computer can do this, given the proper instructions. But the Google brain wasn't given any instructions. After seeing so many similar arrangements of pixels in image after image, it automatically deduced—of its own accord—that those particular patterns must be important, and so it prioritized them. A senior Google engineer who helped to design the software explained to the *New York Times*, "We never told it during the training, 'This is a cat.' It basically invented the concept of a cat."[29]

It is premature at this stage to make meaningful comparisons between the human brain and Google's version. For starters, Google only utilized one billion connections whereas the human brain uses over one trillion. Also, we are yet to understand the mechanism by which the brain neurons actually "learn." What the Google brain did demonstrate is that a machine can, to some extent, come to its own deductions about random information without being programmed. The ability to identify a cat, however, is unlikely to pass the Turing Test, which Alan Turing designed to test the humanness of a computer. But, as the Cato Institute's Julian Sanchez wrote on Twitter, it may well pass the "Purring Test."[30]

The cat experiment is just a foretaste of Google's ambitious program to become the leading developer of artificial intelligence. Over the past few years it has purchased numerous AI businesses, including Nest Labs, Boston Dynamics, and the secretive cutting edge British start-up DeepMind—with the aim of creating a "Manhattan project of AI" that will revolutionize the industry.

Meanwhile, the race is on—in corporations, think tanks, the military and academia—to develop more advanced forms of AI. For example, computer engineers at Sony's Computer Science Laboratory in Paris have designed machines that can compose music. According to Marcus du Sautoy, the Simonyi Professor for the Public Understanding of Science and a professor of mathematics at the University of Oxford:

"One of the big successes has been to produce a machine that can do jazz improvisation live with human players. The result has surprised those who have trained for years to achieve such a facility. . . . What's extraordinary is that the programs in these machines are learning and changing and evolving so that very soon the programmer no longer has a clear idea of how the results are being achieved and what it is likely to do next. It is this element of getting more out than you put in that represents something approaching emerging intelligence."[31]

## Democratizing intelligence

Over time, as scientists learn to integrate artificial intelligence with our own brains, we will take a big leap forward in terms of our cognitive capacities. It will mean that practically anyone can become a geek—even if they were born a dandelion rather than an orchid. The cognitive revolution will democratize intelligence by making it available to more people—provided they are willing and able to pay the price.

There will, of course, be risks with cognitive enhancement—as there are with any new technology. An enhancement may not work for certain people—or worse, it may leave them physically or mentally impaired. As the evolutionary biologist Rob

Brooks observed to me, "The genetic variations in people will ensure that not everyone will respond to a smart pill in an optimum way. There can be unintended consequences." But these concerns are unlikely to discourage people from attempting to join the ranks of the cognitive elite. Nor will they prevent commercial interests from exploiting the growing demand for such products. The cognitive revolution is already well under way—not just in laboratories around the world, but in schools, colleges and workplaces. As discussed earlier, "off label" drugs are now widely used to boost mental performance, as are various other cognitive tools. Moreover, people are using them not just because they want to, but because they feel they have to keep up with their peers.

More broadly, our evolution into a more cerebral society—driven by cognitive-enhancement technologies—offers many advantages. It will enable us to make smarter decisions, which should notionally help with global challenges such as climate change and pandemics. We will become better at building and managing the complex technological infrastructure that underpins our world. Our extra brain power will help us to use our resources more efficiently. Even a 3 percent increase in global IQs, as noted earlier, could generate many economic and social benefits.

The cognitive revolution, however, is not just about making people smarter—or more geek-like. It is a stepping-stone to a bigger transformation confronting humanity. For by expanding our mental capacity—especially through artificial intelligence—we will eventually be able to transcend our biological and environmental limits in ways that will change the course of human destiny. Indeed, this is what unnatural selection is really about—the next phase of our evolution: transcendence.

# Chapter Five

## *Transcendence*

High above the floor of many rainforests there lives a species of orchid that clings to the trees of the canopy. It is not a parasite and does not feed off the trees. It gains its nutrients directly from the air, rain and sun.

This orchid is called an epiphyte—from the Greek *epi* ("above" or "upon") and *physis* ("growth"). It is mostly found in the "cloud forests," which lie at an altitude of between 1,000 and 2,000 meters (3,300 and 6,000 feet). In this fog-laden environment, the epiphyte sucks up moisture through its specialized aerial root system, which has large flat surface areas, and features a spongy epidermis made of dead cells that can absorb humidity.

These "air-breathing" epiphyte orchids have evolved over millions of years to climb up from the dark forest floor in search of sunlight and nutrients. Today there are literally thousands of species, ranging from the ubiquitous dendrobium and phalaenopsis orchids, which can be found throughout India, Asia and Australasia, to extremely rare varieties. According to Rhett Butler, founder of the popular nature site Mongabay.com, there are certain species of epiphytes so rare that they can only be found in microhabitats such as a single Andean valley or a particular canyon in the Guyana Shield of South America.[1]

The story of the epiphyte shows how a species can adapt to, and ultimately transcend, its biological and environmental limitations. This remarkable orchid literally reached for the sky.

We humans have been doing the same thing but on a much grander scale. For by radically reshaping the world in our own image to bring about the Anthropocene, we are moving beyond the biological and physical limits that once confined us. At the forefront of this transformation are the geeks, who, like their botanical brethren, the epiphyte orchids, are reaching for the sky. Their technological breakthroughs and adaptations are helping our species to reinvent itself—cognitively, physically and perhaps even spiritually. Challenges that once seemed insurmountable are now routinely being solved through ingenious new technologies. We now live in an age when even the most debilitating setbacks can be overcome.

This fact was brought home to me recently while flying from Bangkok to London. I was seated next to a robust-looking young man and shortly after take-off I noticed there was something odd about my neighbor's lower left leg. It was artificial. We got chatting and he explained to me that he was a British soldier, and that his leg had been blown off in Afghanistan, when his platoon was caught in a minefield in Helmand province. He had been trying to rescue a seriously injured comrade when he jumped over a dried-out riverbed and landed on the mine. During the violent chaos, the platoon's corporal was killed, and three other men lost their legs in the unmarked minefield.

For his extraordinary bravery, Stuart Pearson received the Queen's Gallantry Medal. He also won a substantial damages claim against the Ministry of Defence, for its failure to provide winches for the Chinook helicopter that was flying over the minefield, and which, if equipped with the winches, could have saved the men.

I asked Pearson how he was coping with the loss of his leg. He responded that he was "just grateful to be alive, considering the circumstances," and he seemed more concerned about the fate of his comrades. He then rolled up his trouser leg to show me his prosthetic limb—a metallic structure topped by a composite plastic knee joint. I suggested it looked like one of those bionic limbs out of a science fiction movie. He explained that his was not the latest version, "There's a new type that lets you walk more naturally because it has little sensors that mirror the action of your good leg. But it's very expensive—about £100,000—so I will use some of the money from the MoD payout to get it."

A few years later, in 2013, Pearson—the trained parachutist— told me that he was jumping out of planes again. Together with a group of fellow ex-servicemen amputees, who had served in such places as the Falklands, Bosnia, Iraq and Afghanistan, he had helped to establish Britain's first disabled skydiving team. Between the seven of them, Pearson calculated, "We have just seven legs and thirteen eyes."[2] The story of Pearson and his skydiving colleagues is a salutary tale of technological progress. If they had been born decades earlier, their disabilities would have confined them to a life of immobility. But nowadays, they can resume active lives again.

Eventually, it is anticipated, wearers of artificial limbs will be able to feel sensations through their robotic fingertips and toes. Similarly, blind people could regain some of their sight through a combination of bioelectronics and retinal implants, which takes video signals from a tiny camera mounted on a pair of eyeglasses and sends them via the retinal nerves to the brain's cortex. Other recent advances in functional restoration include artificial valves, lungs and even hearts. It might even be possible, scientists believe, to "bioprint" these organs and tissues by using human cells instead of ink; no more waiting for a donor. In fact, bioprinting of skin and blood vessels is now a reality.

The point is that these advances won't just stop at restoring lost functionality. They will eventually be used to improve performance by making people stronger, faster and more agile. It is also inevitable that these physical enhancements will be combined with the types of cognitive enhancements discussed earlier, to create more capable human beings. In other words, the various enhancement technologies will begin to converge.

## Convergence

Until quite recently, different human enhancement technologies have developed in separate streams. That is, medical engineering firms have focused on neural implants, prosthetics manufacturers on artificial limbs, pharmaceutical companies on pills, and so on. Although there has been some sharing of technologies, generally speaking, innovation has been industry-specific. Increasingly, however, each new innovation is now being treated like a piece of Lego that can be readily transposed from one industry to another and recombined to create more innovative products. This is called "technological convergence." A prime example of this convergence process is the modern smartphone, which incorporates an array of technologies that were originally developed for other purposes, such as telephony, GPS, Internet, music players, cameras, alarm clocks, barcode readers, radios and so on.

Technological convergence greatly speeds up the development process by forcing applications to work together synergistically, and encouraging manufacturers to spur each other on to greater innovation by competition. Convergence is a force multiplier and accelerator. We see it at work in the military, where a lethal array of ballistics, robotics, computers, sensors and communications technologies have converged in weapons like the Predator drone, a combat system so advanced that it was considered the realm of science fiction just a few years ago. In the civilian world, too, technological convergence is transforming many aspects of our lives. Car companies, for example, are pouring huge amounts of money into telematics—telecommunications and informatics—to develop vehicles that can almost drive themselves, and which will join convoys of other "intelligent" cars that can follow each other along the freeway.

Now this same technological convergence process is driving human enhancement. Manufacturers of neural implants, for example, are looking to integrate the latest advances in pharmacology and computer science in order to optimize their products. Pharmaceutical companies are utilizing genetic engineering to maximize the potency of their existing drugs and to develop new ones, such as smart pills. Prosthetic limb manufacturers now incorporate an array of disparate technologies, which will eventually revolutionize our concept of mobility. We are on the verge of a human enhancement boom, propelled by converging technologies. Ray Kurzweil, the American inventor and futurist, called this phenomenon the law of accelerating returns. "Through thick and thin, war and peace, boom times and recessions, technological progress happens exponentially, not linearly," he wrote in his book *The Singularity Is Near: When Humans Transcend Biology*.[3]

Indeed, the human-enhancement industry has now developed so much momentum that it is only a matter of time before we see the emergence of the first radically enhanced people. These modified humans will be the personification of unnatural selection.

Which brings us to a difficult question: when a person has been so transformed by technology, where does the human being end and the machine begin?

## Transhumanism

There is a term for the blurring of human biology and technology. It is called transhumanism (abbreviated to H+ or h+), and was first alluded to in the book *The Future of Man*, by French philosopher Pierre Teilhard de Chardin. He wrote: "Liberty, that is to say, the chance offered to every man (by removing obstacles and placing the appropriate means at his disposal) of 'trans-humanizing' himself by developing his potentialities to the fullest extent."[4]

This idea was expanded upon by the English evolutionary biologist Julian Huxley in his 1957 book *New Bottles for New Wine*, in which he wrote:

> The human species can, if it wishes, transcend itself—not just sporadically, an individual here in one way, an individual there in another way, but in its entirety, as humanity. We need a name for this new belief. Perhaps transhumanism will serve: man remaining man, but transcending himself, by realising new possibilities of and for his human nature. "I believe in transhumanism": once there are enough people who can truly say that, the human species will be on

the threshold of a new kind of existence, as different from ours as ours is from that of Peking man [one of the first examples of *Homo erectus*]. It will at last be consciously fulfilling its real destiny.[5]

It wasn't until the 1960s, however, that the transhumanist movement really captured the public imagination. This was a period of social and cultural upheaval in which people were looking for new ideas—not just countercultural ones—that could explain man's place in the universe. This curiosity was heightened by the Apollo moon landings (1969 to 1972), which suggested man's destiny was now being propelled by technological forces.

An intriguing and influential figure during this period was the philosopher and futurist FM-2030, who wrote *Are You a Transhuman?* and taught at the New School in New York City and at the University of California (Los Angeles).[6] The son of an Iranian diplomat, FM-2030 changed his name from Fereidoun M. Esfandiary to break free of conventional naming practices, which he saw as a relic of humankind's tribalistic past. "Conventional names define a person's past: ancestry, ethnicity, nationality, religion," he wrote in his book. "I am not who I was ten years ago and certainly not who I will be in twenty years. . . . The name 2030 reflects my conviction that the years around 2030 will be a magical time."

FM-2030 was fascinated by the influence of technology on people's lifestyles and society. He believed that mankind was evolving into a more hybrid—or transhumanist—form that would incorporate physical and mental augmentations, including prostheses and reconstructive surgery. In the late 1970s he also accurately predicted the advent of in-vitro fertilization, teleconferencing, telemedicine, teleshopping and the correcting of genetic flaws. "I am a 21st century person who was accidentally launched in the 20th. I have a deep nostalgia for the future," he wrote. When he died on July 8, 2000, he was placed in cryonic suspension at the Alcor Life Extension Foundation in Scottsdale, Arizona. Many of FM-2030's ideas were carried forward by people such as the British philosopher Max More, who articulated the first principles of a transhumanist philosophy in the early 1990s and helped pave the way for the world transhumanist movement. "Let us blast out of our old forms, our ignorance, our weakness, and our mortality. The future belongs to posthumanity," More wrote in his essay "On Becoming Posthuman."

Today transhumanism has become a common theme in both science fiction and popular culture—and one which is affirmed by each new genetic or technological breakthrough. We are set on a path to transform ourselves—physically, and most importantly, cognitively. The human mind is about to become more capable and powerful. We are not just talking here about modest increases in intelligence or creating more geeks. The transhumanist revolution offers the potential to create a race of super-geeks—people with extraordinary cognitive capabilities (and physical ones too).

The advent of transhumanism is not just another historical transition like evolving from a hunter-gatherer society into an agricultural one, or from the pre-industrial to the industrial age. During these transitions, we humans did not change. We just did different kinds of work, made better machines, and created a larger imprint on the Earth. But we remained essentially human in mind, body and spirit. Transhumanism, on the other hand, confronts us with a monumental shift in the direction of our own evolution—from natural to unnatural. It is the inevitable consequence of unnatural selection.

Many philosophers and scientists are strong supporters of the transhumanist movement, and suggest that—regardless of what we think about it—it is going to happen anyway. So we might as well wave the traffic in the direction it is already heading. Ray Kurzweil believes that transhumanism is the next step in our development as a species, and that enhancement technologies can greatly improve our intellectual, physical and psychological capacities. He has suggested that these "liberation" technologies designed to free us from biological constraints might even enable humanity to realize its age-old dream of immortality. Another influential transhumanist, Ronald Bailey, the author of *Liberation Biology: The Scientific and Moral Case for the Biotech Revolution*, wrote in his book that "the [transhumanist] movement epitomizes the most daring, courageous, imaginative, and idealistic aspirations of humanity."[7]

By freeing us from the blind forces of natural selection, transhumanists argue, the capacity to "self-evolve" will give us an evolutionary advantage, because it means we can plan ahead and make proactive adaptations in response to anticipated threats or environmental changes. It is evolution with foresight. According to Professor John Harris, a professor of bioethics at the University of Manchester, advances in genetics and human fertilization technology "may help us avoid going extinct due to our vulnerabilities and instead enable us to choose (or become) our successors as a species."[8] Writing in the *Times*, he argued that enhancement evolution will ensure that our evolutionary process will be directed by the desires and ideas of intelligent beings. Our human diversity will "no longer be generated by random mutations but by deliberate interventions, which can be very complex (e.g. synthetic biology)." He suggested it might even lead to "the emergence of a new species that will initially live alongside us and eventually may entirely replace humankind."

Like many transhumanists, Harris believes that it is only natural that we take charge of our evolutionary destiny, and that we have actually been striving to do so for a long time. "With things like books or glasses, let alone pacemakers or prosthetics, we are already enhanced," he wrote. "Enhancing ourselves is just what we do as human beings. So not to seek to transcend our humanity would be a denial of our humanity."

# Chapter Six

# *Backlash*

## *The Wings of Icarus*

The theme of human enhancement was examined in an exhibition called "Superhuman," which opened in July 2012 at the Wellcome Collection Gallery in London. The exhibition, dedicated to exploring the connections between medicine, life and art, showcased a variety of human enhancement technologies, and included contributions from artists such as Matthew Barney, and scientists, ethicists and commentators. Exhibits included a prosthetic toe from around 600 BC; artificial legs from Victorian times; Nike sports shoes from the 1970s; a state-of-the-art i-Limb Ultra prosthetic hand; a packet of Viagra; and even a strap-on penis. Super-enhanced heroes, such as Iron Man and Spiderman, were also represented in the form of original artwork from their comics.[1]

For many visitors to "Superhuman," perhaps one of the most indelible images was the statue of Icarus that greeted them at the entrance. This is the young man who in Greek mythology was imprisoned with his father, Daedalus. He subsequently tried to escape by using wings made of feathers and wax. As he was about to take off his father warned him, "Keep to the middle way. Don't dip into the sea or fly too close to the sun."

But Icarus ignored his father's warning and soared higher and higher until the sun melted the wax holding his wings together, and he disappeared into the sea.

Like Icarus we—as a species—are also creating wings to help us soar into the future and achieve our transhumanist destiny. But as with Daedalus, not everyone is confident that these enhancements will guarantee us a safe flight. For as we shall see, the high-tech tools of unnatural selection carry some big risks.

When Francis Fukuyama, the American political scientist and author of *The End of History and the Last Man*, and member of the US President's Council on Bioethics, was asked—along with eight other prominent policy intellectuals—by *Foreign Policy*, "What ideas, if embraced, would pose the greatest threat to the welfare of humanity?" he nominated transhumanism.[2] "It is tempting to dismiss transhumanists as some sort of odd cult," he wrote. "These are the type of people who want to freeze themselves cryogenically in hopes of being revived in a future age."

But despite their place on the intellectual fringe, Fukuyama has suggested we should take them seriously because they are committed to radically changing human life as we know it. Fukuyama believes that transhumanism of a sort is already implicit in much of the research agenda of contemporary biomedicine. He cites new procedures and emerging technologies—such as drugs to boost muscle mass, mood-altering pills, and gene therapy—as initiatives that could be used to enhance the human species. He warns, however, that the intellectual or moral threat these developments represent is by degree not always easy to identify. Although "society is unlikely to fall suddenly under the spell of the transhumanist worldview . . . it is very possible that we will nibble at biotechnology's tempting offerings without realizing that they come at a frightful moral cost."

Like a number of philosophers, Fukuyama fears that the first casualty of transhumanism might be equality. This is because underlying the idea of equality is the belief that we all possess a human essence that dwarfs our differences in skin color, beauty and even intelligence. Modifying that essence is the core of the transhumanist project. "If we start transforming ourselves into something superior," Fukuyama asked, "what rights will these enhanced creatures claim, and what rights will they possess when compared to those left behind?" He worries that for citizens of the world's poorest countries—for whom biotechnology's marvels likely will be out of reach—the threat to the idea of equality will be even more menacing.

We have already seen how geeks with inborn cognitive advantages have achieved extraordinary levels of wealth, and, some would argue, other big privileges too. A transhumanist version of a geek would be many times more capable and formidable.

Another concern about transhumanism is the dilemma of unintended consequences. Human biology is so infinitesimally complex that if we deliberately modify a trait, it is difficult to anticipate the consequences. Unlike natural selection, which is a long-term process that ensures that we evolve in a biologically sustainable way, unnatural selection operates almost immediately. There is no trial period to weed out bugs in the system. Gerald Joyce, a scientist at the Scripps Research Institute, writing in the publication *Science*, warned that: "As one contemplates all the alternative life forms that might be possible with [synthetic genetics], Arthur C. Clarke's novel *2010: Odyssey Two* comes to mind, in which HAL the computer tells humanity, 'all these worlds are yours,' but warns—'except [Jupiter's moon] Europa, attempt no landings there.' Synthetic biologists are beginning to frolic on the worlds of alternative genetics but must not tread into areas that have the potential to harm our biology."[3]

For many people the biggest technological threat is not synthetic biology, but the kind of artificial intelligence discussed earlier. For the possibility of a machine that will be able to "learn" is both thrilling and chilling. Thrilling because it could eventually surpass our own intelligence and help us solve many big problems that our society grapples with; chilling because it could decide that we humans are, for example, a threat to other life forms that it deems to be more necessary to support the planet's biosphere, and that therefore some culling is required. Variations of this scenario have formed the storylines of the dystopian films *Terminator* (and its successors) and *The Matrix*, and books such as Hans Moravec's *Robot: Mere Machine to Transcendent Mind*, which predicts that machines will attain human levels of intelligence by the year 2040 and eventually threaten us with extinction.

It is not just futurists and science fiction writers who believe that artificial intelligence will triumph over natural intelligence. Steve Wozniak, cofounder of Apple Computers told an audience on the Gold Coast of Queensland in 2012 that machines have already "won the war and the human race is destined to become little more than house pets. . . . We're already creating the superior beings, I think we lost the battle to the machines long ago. Every time we create new technology we're creating stuff to do the work we used to do and we're making ourselves less meaningful, less relevant."[4]

When he started Apple with Steve Jobs, he said, he never thought a computer would be powerful enough to hold an entire song, "and today we can fit 50 movies on a little disc in an iPhone. . . . Once we have machines doing our high-level thinking, there's so little need for ourselves and you can't ever undo it—you can never turn them off."

Another prominent computer luminary-turned-bellringer is Jaan Tallinn, who cofounded Skype and is a board member of the Lifeboat Foundation, a nonprofit organization dedicated to minimizing the risks of technologies that include genetic engineering, robotics and artificial intelligence. In July 2012, he told an audience at the University of Sydney that the world is witnessing an "intelligence explosion," and that neuroscience is advancing in leaps and bounds to the point where scientists might be able to replicate the human brain by the mid-twenty-first century. "This is not science fiction, this thing is not apocalyptic religion—this thing is something that needs serious consideration."[5]

Tallinn believes that the real danger will come when computers can do their own programming, because then they will basically take over technological progress. The question then will be, he has said, how can we control something that can actually reprogram itself? It might not be possible. "At some point, humans may no longer be the smartest species," he warns. "It really sucks to be the number two intelligent species on this planet. Just ask gorillas. They will go extinct, and the reason why they will go extinct is not that humans are actively conspiring against the gorillas, it's that we as the dominant species are rearranging the environment. The planet used to produce forests but now it's producing cities. We don't want a super intelligence to do terraforming projects and start changing the atmosphere or soil or whatever."

History tells us that a biological species almost never survives encounters with superior competitors. The worst case scenario, says Tallinn, is not that machines rise up to destroy humanity, but that some form of super intelligence decides it knows what is best for us and keeps us alive under a kind of artificial totalitarianism.

In his seminal book *The Age of Spiritual Machines*, Ray Kurzweil predicted that machines with human-like intelligence would eventually merge with humankind to become one and the same. This convergence process is known as the "singularity," a term borrowed from astrophysics that refers to a point in space-time at which the normal rules of physics no longer apply. According to this scenario, the future won't be characterized by a conflict between "we humans" and "those robots"—or a malevolent HAL computer taking over the world. Rather, as we merge into the singularity, human consciousness will be downloaded into machines, and vice versa. Kurzweil believes that this process may happen so gradually, perhaps over hundreds or even thousands of years, that we won't be sufficiently aware of it enough to care.

Indeed, this process is under way. We are already "uploading" much of our identity—our thoughts, feelings, emotions—into computers, which, in turn, form the basis of our interactions with the world. We now experience so much of our lives through a screen—on a smartphone, iPad, TV or computer—that we begin to perceive this artificial interface as reality itself. Or, as the Zen masters would say, to mistake the finger pointing at the moon, for the moon.

## Losing touch

Our new technology encourages us to live in our heads—as cerebral beings—and to cut ourselves off from our physical selves. In Facebook terms, it wants us to "de-friend" our bodies. So, over time, we begin to resemble a character called Mr. Duffy in

James Joyce's book *The Dubliners*, who is described as "living a short distance from his body."

But we humans are much more than our "minds." We have evolved over millions of years and are deeply grounded in ancient somatic (bodily) processes which regulate our well-being. It is easy to take these for granted in an age of technological wizardry—but we do so at our peril. For these primordial physical aspects of our being are integral to our identity as humans. We cannot afford to lose touch with them.

We learned earlier with the gamers in "The Rise of the Geek" how dangerous it can be for people to disconnect from their bodies. Some gamers become so immersed in their virtual interface they are completely oblivious to their physical needs, leading to death. It is not just extreme gamers, however, who are at risk from being alienated from their physical bodies. Anyone who uses lots of technology is vulnerable, and in these days this means most of us.

The reason why the mind–body disconnection is so dangerous is because our experiences and emotions are processed at the physical level, not just at the intellectual level. When this natural process is blocked or circumvented, it can cause a host of psychological and physical ailments including stress, anxiety, trauma, depression immune dysfunction, and in extreme cases, even death.

Dr. Peter A. Levine, a renowned expert on the treatment of post-traumatic stress, and a former NASA consultant, has said that we cannot deal with life's major experiences in our heads—they must be processed, transmuted and ultimately released through the body. He points to the example of how animals physically shake out the effects of a stressful event through trembling, whereas humans tend to internalize those effects, or distract themselves. What we should be doing, he advised in his book *Waking the Tiger: Healing Trauma*, is getting in touch with our physical bodily sensations, and really feeling them, so that the underlying tensions and stresses can be discharged naturally.[6] But this is easier said than done, he wrote, because so much of our lives—and indeed our survival—today depends on our thinking ability rather than our physical capacity to respond:

"Consequently, most of us have become separated from our natural, instinctual selves—in particular, the part of us that can proudly, not disparagingly, be called animal. Regardless of how we view ourselves, in the most basic sense we literally are human animals. The fundamental challenges we face today have come about relatively quickly, but our nervous systems have been much slower to change. . . . Without easy access to the resources of this primitive, instinctual self, humans alienate their bodies from their souls."

Levine told me that he is particularly concerned about generations of young people who "use technology to keep in touch with dozens of people every hour in cyber-relationships, while authentic face-to-face engagement is clearly on an apocalyptic wane."

The ability to connect with our interior state is sometimes called the "felt sense," a term coined by the American psychotherapist Eugene Gendlin, who discovered that his patients healed much faster when they intuitively focused on very subtle and vague internal bodily feelings during their therapy. A similar approach is espoused by practitioners of mindfulness meditation, which encourages people to "befriend their bodies" by noticing their physical sensations in a non-judgmental way. Dr. Jon Kabat-Zinn, who founded the mindfulness-based Stress

Reduction Clinic at the University of Massachusetts Medical School, wrote in his book *Wherever You Go, There You Are*, that "profound healing and transformation can come from cultivating mindful awareness of our bodies, even for those suffering chronic illness. . . . There are very few things that don't go better than when we are fully embodied and present."[7]

Unfortunately, our techno-centric society is pushing us in the opposite direction—to be less embodied and less present. Many people nowadays are so busy multitasking—responding to emails, reading Twitter, updating Facebook—that their attention is everywhere and nowhere at the same time. They are rarely fully alert to the here and now, let alone to their bodies. It is as if they have abdicated the physical realm in favor of a cyber-existence—and have become prototypes of a transhumanist future.

These are early days yet, of course, but the direction of the cognitive revolution is becoming increasingly clear. Our quest to become a smarter, more geek-like species may eventually threaten the foundations of our humanity. As we have seen, this has prompted many scientists, philosophers, policymakers and citizens to speak out about the specter of human enhancement. Of course, it is not the first time in history that people have railed against new technology.

## The return of King Ludd

Just over two hundred years ago in Nottinghamshire, England, groups of cloth workers would gather at night under the banner of a mythical figure named King Ludd, who, like Robin Hood, was believed to live in nearby Sherwood Forest. In the darkness, they would rehearse their maneuvers on the local moors, before proceeding to the nearby textile mills where they would smash the new looms that were threatening their livelihoods. These new wide-framed automated looms could be operated by cheap, relatively unskilled labor, resulting in the loss of jobs for many skilled workers—who were fast becoming more casualties of the Industrial Revolution that was sweeping Europe.

The Luddite movement soon spread across the north of England and caused the destruction of many more wool and cotton mills. At the height of the conflict one mill owner was assassinated and the government—fearing the rise of a national movement—mobilized the army to crush the protesters. Soon there were more British soldiers fighting the Luddites than fighting Napoleon on the Iberian Peninsula. In one of the most ferocious incidents, in April 1812, about 2,000 protesters descended on a mill near Manchester resulting in the death of eight protesters and the wounding of many more. Soldiers eventually brought the situation under control and numerous Luddites were brought to trial in 1813, some of whom were executed or sent to the penal colonies in Australia. "Machine breaking" (industrial sabotage) was subsequently made a capital crime, despite protestations by sympathizers such as the poet Lord Byron and the writer Mary Shelley, whose concerns about technology were echoed in her novel *Frankenstein*.

The Luddites weren't so much against modern technology—which they saw as inevitable—as against the way that the new class of factory owners used it as a pretext to drastically cut the wages and conditions of workers. In his book *Writings of the Luddites*, the literary scholar Kevin Binfield pointed to the fact that the Luddites confined their attacks to those manufacturers who used the technology in a "fraudulent and deceitful manner" to bypass standard labor practices.[8] Nevertheless, over time the term "Luddite" has come to mean someone who resists technological change—which, in today's gadget-obsessed world, is seen as being out of step with reality. In light of the ominous scenarios discussed earlier, however, some people are calling for the return of King Ludd—at least in spirit. Like their predecessors, these people are not "against technology" per se, but rather the way it might be used. Specifically, they want safeguards to ensure that the human enhancement industry doesn't spiral out of control, with unintended consequences of gross inequality and the emergence of a new cognitive elite. The trouble is that it is difficult to regulate against new technologies that offer life-enhancing benefits, as we have seen in recent years with stem-cell innovations. Indeed, the only type of government that could exert any meaningful control over the juggernaut of human enhancement would be a totalitarian one. And this would surely end up driving scientists underground where they would be subject to even less supervision. Also, as discussed earlier, many governments are keen to exploit the fruits of the cognitive enhancement evolution for military and security purposes.

The other big regulatory hurdle is that the technology is so new we cannot yet quantify the risks. And legislation cannot be based on speculation.

## The precautionary principle

There is an alternative approach, however, that has been successfully applied to other fields where the science is still evolving, such as climate change. It is called the "precautionary principle," and was developed in 1998 by a group of thirty-five scientists, lawyers, policymakers and environmentalists from the United States, Canada and Europe, who came together to figure out a way to grapple with serious potential threats to society as a whole. The principle applies where "the scientific evidence is insufficient, inconclusive or uncertain and prelimi-nary scientific evaluation indicates that there are reasonable grounds for concern," and therefore that steps should be taken to minimize the potential harm.[9]

The field of human enhancement would certainly meet these criteria, and is therefore a good candidate for applica-tion of the precautionary principle. The challenge will be to ensure the principle is designed in such a way that it is relevant, coherent and effective. Fortunately, a number of thinkers and organizations have given this issue some consideration. The late writer William Safire, for example, was keenly interested in the implications of cognitive enhancement, and suggested that any governing principles or standards must "examine what is right and wrong, good and bad, about the treatment of, perfec-tion of, or unwelcome invasion of and worrisome manipulation of the human brain."[10]

In other words, we should first consider the ethical and moral dimensions, not the scientific ones. For this is the only way to ensure that the technology serves us, rather than the other way around. To this end, a number of organizations have begun work on developing ethical frameworks for the cognitive-enhancement industry and the neuroscience that underpins it. The International Neuroethics Society, for example, has been established "to promote the development and responsi-ble application of neuroscience through interdisciplinary and international research, education, outreach and public engagement for the benefit of people of all nations, ethnicities, and cultures."[11]

Another organization that is pushing for standards is the Wellcome Centre for Neuroethics (which was behind the "Superhuman" exhibition discussed earlier). It was established by the University of Oxford in 2009, to address concerns about the effects that neuroscience and neuro-technologies might have on various aspects of human life.[12] Its research focuses on five key areas: cognitive enhancement; borderline consciousness and severe neurological impairment; free will, responsibility and addiction; the neuroscience of morality and decision-making; and applied neuroethics. The center proposes a similar approach to that advocated by William Safire, which is that any meaningful set of standards must address the following questions:

*Definition*: What constitutes enhancement? How is it related to concepts of disease, therapy and normal functioning? Should we understand enhancement neutrally, in terms of increased function, or in value-laden terms, in terms of promotion of well-being?

*Morality*: Is enhancement morally permissible, even required, or is its pursuit morally hazardous? Are moral objections to enhancement valid and can they be answered? How, if at all, do controversial forms of enhancement differ from current means of improving capacities, such as better nutrition?

*Implications*: To the extent that enhancement is permissible, what would be the social and global effects of widespread use?

To answer these questions, however, we must first grapple with the most perplexing issue of all, which is: what constitutes the human identity?

Currently, much of our sense of being human derives from our physical form—our unique face, body, fingerprints, hair, and even the way we walk and talk. Certainly there are other elements to our identity, such as our names, reputations, occupations and possessions—but these are all subordinate to our physical form. This even applies to highly cerebral types. Think for a moment of one of the most brilliant geeks who ever lived—Albert Einstein—and what first comes to mind? I bet you it is an image of his wizened face topped with that crazy hair.

As we enhance ourselves through the tools of unnatural selection—acquiring cognitive implants, robotic extensions, augmented reality systems, and genetic engineering—our biological identity will become less certain. It will morph into an agglomeration of bionic and genetically modified components. Even the thoughts, beliefs and value systems which define our psychological identity could, in time, be subsumed within the cognitive-enhancement process. Such enhancements might, for example, encourage us to become more rational and less driven by the vicissitudes of our emotions. That is, to be more like *Star Trek*'s cerebral Mr. Spock and less like the emotional Captain James T. Kirk.

It is also likely that human enhancement will be driven by fads and trends, and affect different groups of people in different ways. So for example, when the stock market becomes troubled, a cosmetic neurologist (a plastic surgeon of the brain) might do great business tweaking bankers' amygdalas—the part of the brain that helps mediate emotions—in order to make them more stress resistant. What all this points to is a more malleable and fluid concept of human identity than we have experienced before. Identity will no longer be a fixed concept, but an ephemeral manifestation of our current intention.

Like the gods of old, we will choose how we wish to manifest ourselves according to the situation. We can see inklings of this scenario on the Internet today, where people routinely use avatars to represent themselves online, or in virtual communities such as Second Life.

Although some of the human-enhancement technologies we have discussed might seem a long way off—we do need to grapple with the issue of human identity today. This is because for a neuroethics framework to be effective, it must be underpinned by a clear sense of what a human being is. Otherwise, how would we know what we are trying to protect? Any guidelines for neural implants, for example, must take into consideration how drastically they alter a human identity, which therefore, must first be defined. It is important that this definition not be too restrictive or narrow. Nor should it be based on purely "naturalist" values, so that only non-enhanced people qualify to be true humans.

Perhaps a good place to start is to revisit the work of the Swedish botanist and physician Carl Linnaeus (1707–1778), who set about classifying all the life forms on Earth in his book *Systema Naturae*, published in 1735. He pondered for a while over what to name humans, and first thought our species should be *Homo diurnus*, meaning "man of the day"—which made sense, given our prominent position in the world. On reflection, he decided that we needed a name that emphasized our uniqueness and distinction from the rest of the animal kingdom. He wanted the name to reflect our species' defining attribute, which he saw as being "wisdom." Hence, he coined the Latin-based term *Homo sapiens*, meaning "knowing" man or "wise man."

Linnaeus wasn't the first person to make the case that wisdom is at the core of our identity. The Greek philosophers

Socrates and Plato believed that the love of wisdom (*philo-sophia*) should be the guiding principle of humankind's philosophy; and the Christian philosopher Thomas Aquinas decreed wisdom to be the "father of all virtues." At least four of the world's great religions—Christianity, Islam, Buddhism and Judaism—regard wisdom as a cornerstone of their ethical foundations. Indeed, our survival as a species depends on our making wise judgments.

It might be a good starting point, therefore, to posit that in the context of unnatural selection, any definition of the human being should incorporate "wisdom" as a key element. The litmus-test question for any new human enhancement should be simply: is this really a wise thing to do? This may help ensure that we use the tools of cognitive enhancement, genetic engineering, prosthetics and pharmacology in such a way that we don't sacrifice our essential humanity. For it would be a mistake, as noted earlier, to lose touch with the innate wisdom of our physical and emotional being that has evolved over millions of years, and which has served us so well. We must learn ways to embrace *all* our disparate parts—physical, emotional, spiritual and biomechanical—thus enabling us to become more than the sum of our parts, so that no matter how enhanced and geek-like we become, we are able to reconnect with the "wholeness" of our being. That is, to be truly human, no matter what the future brings forth.

## Valhalla (summary)

Overlooking Lake Zurich in Switzerland, there is a majestic old home called Villa Wesendonck. It was built in 1857 for a wealthy German industrialist, Otto Wesendonck, and soon became a meeting place for the cultural elite of Zurich and else-where. A regular visitor to the villa was the composer Richard Wagner, who eventually moved in with the Wesendoncks and became infatuated with Otto's beautiful young wife, Mathilde.

It is easy to imagine, while walking the grounds of this magical place, why Wagner delighted in being here with his new muse. Lush gardens and cobble-stoned lanes surround the neoclassical villa, and on a clear day you can glimpse the Swiss Alps in the distance through the giant oaks, maples and fir trees. In springtime, the gardens burst forth with flowering blooms and the sounds of insects and birds, including the odd woodpecker rattling against the hardwoods. The buildings and grounds have been faithfully preserved in their original style, and prudently screened from the neighboring urban landscape by a ring of dense foliage. Indeed, it is such a timeless place, that you could readily believe that nothing much has changed since Wagner and Mathilde cavorted here under the trees over a hundred and fifty years ago.

This frozen moment is, however, an illusion.

A few meters below the manicured lawns, there is a high-tech structure of cavernous proportions, where the temperature never deviates by more than two degrees and the loudest noise is the whoosh of lift doors delivering fresh visitors from the surface world. Luminous ceilings bathe each floor in artificial light, and twenty-four-hour motion-detection sensors monitor the entire area. A network of fiber-optic cables runs below the cobble-stoned lanes to connect the building to specialized data-bases around the globe.

This underground edifice is part of the Rietberg Museum, and it is one of the most technically advanced structures of its kind in the world—a fine tribute to European engineering. It was

designed by the architects Alfred Grazioli and Adolf Krischa-
nitz to house a vast collection of Eastern, Asian and Oceanic
art, which is laid out on a grand scale. But unless you were actually looking for it, you would hardly know it was there. Only
the entrance—a discreet emerald-green glass cube—is visible
from the outside.[13] The building is so fully integrated into the
existing landscape that you could spend hours frolicking in the
bucolic grounds of Villa Wesendonck and be completely oblivious to its existence.

In many ways Villa Wesendonck is a metaphor for our world
today. On the surface, things appear to be relatively unchanged,
but dig a little deeper and a whole new vista opens up.

In this book, we have explored some of the changes that
are occurring beneath our feet—often out of sight—as we step
further into the Anthropocene. At first we feel these changes
as slight tremors that are barely noticed. But eventually—and
inevitably—the ground will begin to move and shake, and as
it does the beliefs and certainties that once sustained us will be
swept aside. Our world is undergoing a seismic shift—geologically, environmentally and especially technologically.

In Richard Wagner's acclaimed opera *The Ring Cycle*,
which he partially wrote at Villa Wesendonck—there is a magical piece of technology called the Ring, which confers on its
holder the power to rule the world. During the course of this
tumultuous opera, which takes place over four consecutive
nights, various people, gods and mythical creatures plot and
scheme to gain control of the Ring. They want to access its infinite power. Eventually, after an inter-generational struggle, the
Ring falls into the hands of the Rhine maidens, who take it to
their city of Valhalla, which is subsequently destroyed by the
power of the Ring.

When George Bernard Shaw reviewed *The Ring Cycle* in his book *The Perfect Wagnerite*, he said Wagner's opera was a critique on the power and dangers of industrialization (technology). He wrote, "Only those of wider consciousness can follow [this opera] breathlessly, seeing in it the whole tragedy of human history and the whole horror of the dilemmas from which the world is shrinking today."[14]

Whether or not Shaw was correct about Wagner's intention, *The Ring Cycle* certainly has something important to tell us about the new epoch we have entered. For the quest to control the world through technology lies at the heart of the Anthropocene—the Age of Man. It has become a key driver of our collective unconscious. Although we embarked on this journey a long time ago—when our hominid ancestors shaped their first stone tools—it is only very recently that we started to shape the environment itself on a global scale, and will eventually be able to shape our own evolutionary destiny.

In Wagnerian terms, we—and especially the geeks—now have the Ring in our hands. Whether we will use it wisely is another matter. For as noted throughout this book, the technologies of the Anthropocene present us with opportunities and perils in equal measure. The *New Yorker* music critic Alex Ross once described Wagner's music as being "marked by a constant tension between a will to power and a willingness to give up."[15] We face a similar tension today, for we have become so addicted to our technology that—like any addict—it is difficult for us to give it up or even to loosen our grip.

This is likely to be particularly true of the cognitive-enhancement technologies, such as smart pills and emerging neuro-implants. Once a person has experienced what it is like to be brighter, sharper and on the ball—they will probably wish to be like this all the time. Like biblical Adam, they will have tasted the "fruit of knowledge" and want more. Many people will come to believe they have no choice but to join the cognitive arms race of the twenty-first century. Popping a pill is a small price to pay for a shortcut to geek-dom.

But there's the thing: by focusing on cognitive development and putting it on a pedestal, we are in danger of mistaking intelligence for wisdom, which, as discussed earlier, is the hallmark of our humanity. These are not the same thing. Many of the smartest people who ever lived were not particularly wise, and vice versa. Indeed, some of the world's most tyrannical monsters and blunderers have been acutely bright people, with great powers of perception and cognitive dexterity. They were the geeks of their day. Yet the very qualities that made them so intelligent also enabled them to justify to themselves almost any belief or action. Studies by cognitive researchers have consistently demonstrated that acutely bright people can be particularly vulnerable to having "blind spots" or biases in their thinking. According to a recent study conducted by Richard F. West, Professor Emeritus at the Department of Graduate Psychology at James Madison University, intelligent people are better at convincing themselves of what they want to believe, regardless of any inconsistencies in their story. In other words, fooling themselves. When Dr. Michael Shermer, author of *The Believing Brain*, was asked why smart people believe in weird things like creationism, ghosts and (as with Sir Arthur Conan Doyle) fairies, he responded, "Smart people are very good at rationalizing things they came to believe for non-smart reasons."[16] That's the thing about intelligence; it doesn't necessarily make you right but it can make you better at thinking you are.

Why does all this matter? Because the rise of the geeks—and geek culture—means that our society now reveres smart people, perhaps more than at any other time in its history. The celebrities of our age are increasingly "geeks"—not explorers, politicians, military leaders, industrialists, conquerors, religious leaders or philosophers. Today's celebrity is more likely to be a bright young person who designs a popular software program and becomes a billionaire by the time he or she is twenty-five years old. Or they could be a financial quantitative analyst (a quant), who creates a new form of derivative that transforms the financial markets. It is this type of person—the geek—who now graces the front pages of magazines and whose counsel is sought by prime ministers, presidents and think tanks. It is their ideas that inevitably percolate through society and influence its values and direction. This means they have an inordinate influence on important issues such as the cognitive revolution and the emergence of a transhumanists society. Indeed, they are now in the driving seat of human evolution.

Yet, as we have seen, the geeks may be bright, but they are not necessarily wise. Recall that the global financial crisis—which still reverberates to this day—was largely caused by complex derivatives trading programs designed by quants. Similarly, one day in the future, if we should ever suffer at the hands of a rogue artificial intelligence program, we can be sure that a geek will have designed it.

The premise of this book, as stated on the cover, is that the geeks will inherit the Earth. Indeed, in many respects they already have. They are certainly among the most influential and successful groups in the world today. Their products, ideas and actions permeate our lives to a far greater extent than the manufactured goods of old industrialists ever did. This is because their products are not like cars or washing machines; they engage our minds and encourage us to mold our thinking processes to match their operating software. We experience the world through their virtual interfaces, and interact with it through their apps. In effect, the geeks have already laid claim to a large proportion of our cognitive terrain, only we don't recognize it because it has occurred through a process of technological osmosis. As Sergey Brin explained when promoting his smart glasses, "We want Google to be the third half of your brain."[17]

Despite this cognitive takeover, most of us believe that it is we who are calling the shots. With our gadgets in hand, empowered by so much information at our fingertips, we feel like virtual gods at the center of a pre-Copernican universe. The reality is, however, that we exist in a digital bubble and our sense of omnipotence is an illusion.

Every so often this bubble bursts, and we are abruptly reminded of our real place in the greater scheme of things. Such reminders often manifest as natural calamities—an earthquake, hurricane or tsunami—that shatter our sense of control. Who can forget, for example, the images of the tsunami that struck in 2004, killing nearly a quarter of a million people throughout South-East Asia? Entire communities were wiped off the map. Then came the 2011 Japanese tsunami that devastated a number of cities in Miyagi Prefecture, killed thousands of people, and triggered a nuclear disaster at the Fukushima Daiichi nuclear power plant.

Fortunately, not all reminders of our human vulnerability are so devastating. There have been times when we have been afforded a more subtle and contemplative glimpse into the nature of our place in the universe.

# A pale blue dot

One of these occasions occurred on February 14 (Valentine's Day) in 1990, when the space probe Voyager I reached the edge of our solar system after travelling for many years. It was now over six billion kilometers from Earth—farther than any man-made object had ever ventured. From this distant position the probe took a remote-controlled picture of Earth and beamed it back to NASA—a transmission that took nearly five and a half hours travelling at the speed of light.

This extraordinary image showed Earth as a tiny pixel-sized speck—a pale blue dot—surrounded by infinite blackness. Never before had our planet looked so inconsequential and insignificant. When the cosmologist Carl Sagan—who had originally proposed that Voyager I take this photo—was able to see it, he was moved to write: "The Earth is a very small stage in a vast cosmic arena . . . . Our posturings, our imagined self-importance, the delusion that we have some privileged position in the Universe, are challenged by this point of pale light. Our planet is a lonely speck in the great enveloping cosmic dark. In our obscurity, in all this vastness, there is no hint that help will come from elsewhere to save us from ourselves."[18]

As we venture further into the Anthropocene, we would be well advised to heed Sagan's words. For it is indeed up to us to save ourselves, especially now that we are opening up the Pandora's box of cognitive enhancement, synthetic biology and artificial intelligence. These tools of unnatural selection have the potential to fundamentally alter the course of human evolution, and, for reasons outlined earlier, may even threaten our survival as a species. Like the Rhine maidens of Wagner's opera, we have the Ring in our hands and are heading for the gates of Valhalla.

Whether we survive or not depends largely on whether we use this new technology for selfish reasons or unselfish. That is, do we focus on enhancing ourselves at the expense of others, or do we take a wider—societal—view? This question is critical because there is now strong anthropological evidence— gathered over the past twenty years—that humans are not primarily driven by the "selfish gene" concept popularized by evolutionary biologists in the footsteps of Richard Dawkins. Rather, if this evidence is correct, we live or die according to our ability to coexist with each other, and to have empathy for others. Although there is always a tension between selfish behavior and altruistic behavior, over time we have learned— often through violent conflict—that groups of altruistic individuals will beat groups of selfish individuals. This is because altruistic behavior enables people to cooperate better, and be more than the sum of their parts. In other words, they can become a society.

It is our ability to cooperate socially—and altruistically— that has enabled the rise of *Homo sapiens*. Importantly, this altruism doesn't just extend to our kin, but more broadly to our tribe or group through what is known as "multi-level selection." The Harvard biologist Edward O. Wilson, author of *The Social Conquest of Earth*, explained in his book that:

"[T]he evolutionary products of group selected behaviors [are] so completely a part of the human condition that we are prone to regard them as fixtures of nature, like air and water. They are instead idiosyncratic traits of our species. Among them is the intense, obsessive interest of people in other people, which begins in the first days of life as infants."[19]

Wilson suggested that there is an overpowering instinctual urge among people to belong to groups. "To be kept in solitude is to be kept in pain, and put on the road to madness. A person's membership in his group—his tribe—is a large part of his identity." This desire for closeness can be traced back over a million years ago when our pre-human ancestor, *Homo habilis*, gathered around a single site for shelter and learned to cooperate with each other, hunt together, share food, and develop a rudimentary form of social intelligence.

As we venture further into the Anthropocene, the foundations of our altruistic society will be increasingly challenged by human-enhancement technologies. These technologies are likely to develop our individual abilities, rather than our collective—or altruistic—ones. Therefore, we should tread carefully, for as Wilson observed, "To yield completely to the instinctual urgings born from individual selection would dissolve society."

This anthropological truth was brought home to me by my ten-year-old son, who is a keen player of an online video game called Minecraft. This is a tactical game in which players fight each other for supremacy and can "enhance" their online personalities—or avatars—by strengthening their armor and developing superior capabilities. After many weeks, my son had enhanced himself to a point where he felt quite invincible. But he was lured into a trap by a rival and his gang, who subsequently destroyed his online identity. My son was shocked and rather humiliated by the experience; young gamer geeks can take these immersive games very seriously.

"I don't get it," he said. "I was fully enhanced and had the best armor and loads of diamond swords. I was a General 5 [the highest rank] and I could even fly. I was the best I could be. But they still got me." We talked about his experience, going over what had happened in detail. But he was still upset and vowed, rather sulkily, that he wouldn't play Minecraft for a while.

The next day he came to me and announced, "I think I know what happened. I thought that I could do it all myself . . . that if I had all those super powers then no one could beat me. But I didn't realize that I was defeated because I was by myself. I should have had my friends with me."

Later that afternoon, he was back online, stripped of all his enhancements. But this time, instead of going solo he was careful to stay close to his colleagues. Inevitably, another battle ensued with the old rival, but unlike before, my son survived.

"I know what this game is all about now," he declared afterwards.

"What?" I asked.

"It's about cooperation."

E. O. Wilson could not have said it better.

# Acknowledgments

This book would not have been possible without the support of many people. In particular I would like to thank my agent, the brilliant Lyn Tranter, for keeping me on track and for being such a wise counsel and friend. Special thanks to Cal Barksdale at Arcade Publishing for bringing *Unnatural Selection* to the United States and Canada, and to his colleague, Maxim Brown, for greatly improving the manuscript.

I am indebted to many researchers, biologists and scientists who have been generous with their time and observations. In particular I would like to thank Rob Brooks, professor of evolutionary biology at the University of NSW, for greatly enhancing my understanding—and appreciation of—the modern evolutionary synthesis; Dr. Jan Zalasiewicz of the University of Leicester, a leading researcher on the Anthropocene; Dr. Cath Suter, head of the epigenetics laboratory, at the Victor Chang Institute, and her fellow researchers; Madeleine Beekman at the School of Biological Sciences at the University of Sydney; Damian Milton, at the Autism Centre for Educational Research, University of Birmingham; Mike Stuart, Community Manager, at Talk About Autism; Robin Jeffrey, visiting professor at the Institute of South Asian Studies in Singapore; Dr. Peter A. Levine, the pioneering trauma expert and ex-NASA consultant, and Michael Darling for his insights and for being an invaluable sounding board.

On a personal note, I am especially grateful to Lottie Horsman for her love, faith and patience. And to Sue Stafford for being such a wonderful and loving mum to Orlando, and for her ongoing support.

Many others, too, have contributed to this work either directly or indirectly, including my parents, Joanne and Reuben Mendick, Clarence Roeder, Max Fulcher, Jon Kabat-Zinn, Roby Abeles, Terry Collinson, David Hale and Lyric Hale, Lord and Lady Leitch (Sandy and Noelle), Kim McKay, Fareed Zakaria, Barney Greer, John Tranter, Janine Bavin, Diana Hannes, Fiona Harris, Bill Miller, Surinda and D. K. Matai, John Hill, Caroline Morgan, Wallace Dobbin, Jeff Beyer, Dr. Adrian Cohen, Dr. Caroline Sein, Maggi Eckardt, Brian Bona, Tom and Lizi Hill, Lucinda Maguire, Linda and Christoph von Graffenried, Rina Canonica, Denise Shaw, Steven and Elizabeth Kennedy, Bob Stafford, John Dawson, Julian Wolanski, Nikola Sekalic, Lucy Messervy, Frank Fowler, Kris Vail, Lynton and Amy Barber, Peter Harris, Rick Spencer, Dr. David Beales, Holly and Nick Saunders, Brent Roeder, Phyllis Roeder, Ken and Mary Betty Roeder, Beverly and Tim Fowler, Nigel Boarer, Matthew Bridle, Jolyon Carpenter, John Hayes, Effie and Maddy Dunscombe, John Barter, Paki Heisserer, Andrea and Jeffrey Mendick, and my family. And last but not least, Heather Turland and her wonderful team at the Depot Café for keeping me well nourished.

# *Endnotes*

## INTRODUCTION

1 Iain Stewart, *How Earth Shaped Us* (BBC documentary aired February 2010).

2 Two interviews with Rob Brooks, Professor of Evolutionary Biology, University of New South Wales, October 2012.

3 Interview with Dr. Jan Zalasiewicz, University of Leicester. See also Howard Falcon-Lang, "Anthropocene: Have Humans Created a New Geological Age?" (May 2011, www.bbc.co.uk/news/science-environment-13335683. Note: Dr. Zalasiewicz's quote to me was almost identical to his words in the BBC article).

4 Annie Murphy Paul, "The Upside of Dyslexia" (*New York Times*, 4 February 2012, www.nytimes.com/2012/02/05/opinion/sunday/the-upside-of-dyslexia.html?_r=0).

5 David Dobbs, "The Science of Success" (*Atlantic*, 1 December 2009, www.theatlantic.com/magazine/archive/2009/12/the-science-of-success/307761/).

6 Mark Mendick (Roeder) & Julian Wolanski, *How to Beat Space Invaders* (Murray Publishing, 1980).

7 See, for example, "The Spice of Life: An interview with Stephen Jay Gould" (*Leader to Leader*, www.hesselbeininstitute.org/knowledgecenter/journal.aspx?ArticleID=64).

8 Michael Le Page, "Evolution in the Fast Lane: Unnatural selection" (*New Scientist*, 2 April 2011).

See also Le Page's excellent article, "How Humans Are Driving Evolution" (*New Scientist*, www.newscientist.com/article/mg21028101.800unnatural-selection-how-humans-are-driving-evolution.html).

9  Quoted in David Biello, "Culture Speeds up Human Evolution" (*Scientific American*, 10 December 2007, www.scientificamerican.com/article.cfm?id=culture-speeds-up-human-evolution).

10  Quoted in John Cloud, "Why Genes Aren't Destiny" (*Time*, 18 January 2010).

11  Richard Dawkins, *The Blind Watchmaker* (W. W. Norton & Company, September 1997).

CHAPTER ONE:
THE ANTHROPOCENE

1  Brett Evans, "A World of Our Own Making" (*Inside Story*, 17 February 2012, inside.org.au/anthropocene/).

2  Erle C. Ellis, Associate Professor of Geography and Environment at University of Maryland, USA, "Land impact calculation" (ecotope.org/people/ellis/).

See also Gaia Vince, "Earth: Will the Age of Man be Written in Stone?" (BBC, 19 December 2012, www.bbc.com/future/story/20121018-will-age-of-man-be-set-instone).

3  "Welcome to the Anthropocene" (*Economist*, 26 May 2011, www.economist.com/node/18744401).

4  David Shukman, "Carbon Dioxide Passes Symbolic Mark" (BBC, 10 May 2013, www.bbc.co.uk/news/scienceenvironment-22486153).

5  John Vidal & Adam Vaughan, "Arctic Sea Ice Shrinks to Smallest Extent Ever Recorded" (*Guardian*, 14 September 2012, www.guardian.co.uk/environment/2012/sep/14/arctic-sea-icesmallest-extent).

See also Jo Confino, "Climate Change: The truth will out" (*Guardian*, 31 October 2012, www.guardian.co.uk/sustainable-business/climate-change-closer-home-confront-truth).

6  Antonio Stoppani, *Corso di Geologia* (Vol. 1, 1873, reprinted by Nabu Press 6 February 2012).

See actual text at Making the Geologic Now: Responses to material conditions of modern life, geologicnow.com/2_Turpin+Federighi.php

7  Quoted in Brett Evans, "A World of Our Own Making" (*Inside Story*, 17 February 2012, inside.org.au/anthropocene/).

8  "Welcome to the Anthropocene" (C&EN, https://pubs.acs.org/cen/editor/86/8605editor.html).

9  "Greening of the Red Planet" (*Science News*, 26 January 2001, science.nasa.gov/science-news/science-at-nasa/2001/ast26jan_1).

See also "Mars Exploration, Colonization, and Terraforming Links" (National Space Science Data Center, nssdc.gsfc.nasa.gov/planetary/mars/mars_colonize_terraform.html).

10  *Science Express* (10 February 2011).

See also Beth Lebwohl, "Martin Hilbert: All human information, stored on CD, would reach beyond the moon" (*EarthSky*, 10 February 2011, earthsky.org/human-world/martin-hilbert-all-human-information-storedon-cd-would-reach-beyond-the-moon).

11  Niall Ferguson, "World on Wi-Fire" (22 October, 2011, www.niallferguson.com/journalism/finance-economics/world-on-wi).

12  Lena Groeger, "Kevin Kelly's 6 Words for the Modern Internet" (*Wired*, 6 June 2011, www.wired.com/business/2011/06/kevin-kellysinternet-words/).

13  Jill Stark, "*Log In, Tune Out: Is technology driving us crazy?*" (*Sydney Morning Herald*, 14 October 2012).

14  Nicholas Carr, "Is Google Making Us Stupid?" (*Atlantic*, August 2008, www.theatlantic.com/magazine/archive/2008/07/is-google-making-us-stupid/306868/).

15  Marika Dobbin, "Gen Y-fi Caught in Web Fixation" (*Sydney Morning Herald*, 14 February 2013, www.smh.com.au/technology/technology-news/gen-yfi-caught-in-web-fixation20130214-2efgy.html).

16 Aleks Krotoski, "Home: How the Internet has changed our concept of what home is" (*Guardian*, 2 October 2011, www. guardian.co.uk/technology/2011/oct/02/aleks-krotoskiuntangling-web-home).

17 Sir Martin Rees speaking at the Planet Under Pressure conference in London, 2012, www.planetunderpressure2012.net

18 Stewart Brand, "We are as Gods and have to get good at it" (*Edge*, 20 August 2009, www.edge.org/3rd_culture/brand09/brand09_index.html).

## Chapter Two:
## Unnatural selection

1 Richard Turner, "Darwin's moth: "Proof of evolution" (BBC, 4 June 2008, www.bbc.co.uk/manchester/content/articles/2008/06/04/040608_peppered_moth_feature.shtml).

2 Daniel Dennet, *Darwin's Dangerous Idea* (Simon & Schuster, June 1996).

3 Theodosius Dobzhansky, "Nothing in Biology Makes Sense Except in the Light of Evolution" (*American Biology Teacher*, vol. 35, pp 125–29).

See also John C. Avise & Francisco, J. Ayala, "In the Light of Evolution, vol. 1: Adaptation and Complex Design" (Proceedings of the National Academy of Sciences of the United States of America, www.pnas.org/content/104/suppl.1/8563.full#ref-24).

See also Leonid Moroz, "The Devolution of Evolution" (*Scientist*, 1 November 2012, www.the-www.the-scientist.com/?articles.view/articleNo/29333/title/The-Devolution-of-Evolution/).

4 "Princeton University Archives: Evolutionary Psychology" (www.princeton.edu/~achaney/tmve/wiki100k/docs/Evolutionary_psychology.html).

See also Steven Pinker, Werner Kalow, Harold Kalant & Stephen Jay Gould, "Evolutionary Psychology: An exchange" (*New York Review of Books*, 9 October 1997, www.nybooks.

com/articles/archives/1997/oct/09/evolutionary-psychology-an-exchange/?pagination=false).

5 David Dobbs, "The Science of Success" (*Atlantic*, 1 December 2009, www.theatlantic.com/magazine/archive/2009/12/the-science-of-success/307761/).

6 Ian Sample, "Quest for the Connectome: Scientists investigate ways of mapping the brain" (*Guardian*, 7 May 2012).

7 W. Thomas Boyce & Bruce Ellis, "Biological Sensitivity to Context: I. An evolutionary–developmental theory of the origins and functions of stress reactivity" (journals.cambridge.org/action/displayAbstract?fromPage=online&aid=303780).

See also Wray Herbert, "On the Trail of the Orchid Child" (*Scientific American*, 22 November 2011, www.scientificamerican.com/article.cfm?id=on-the-trail-of-the-orchid-child).

See also David Dobbs, "The Science of Success" (*Atlantic*, 1 December 2009, www.theatlantic.com/magazine/archive/2009/12/the-science-ofsuccess/307761/).

8 Wray Herbert, "On the Trail of the Orchid Child" (*Scientific American*, 22 November 2011).

9 Danielle M. Dick et al., "CHRM2, Parental Monitoring, and Adolescent Externalizing Behavior: Evidence for gene-environment interaction" (*Psychological Science*, 24 March 2011).

10 G. M. Slavich & S. W. Cole, "The Emerging Field of Human Social Genomics" (*Clinical Psychological Science*, 2013).

11 Daniel Dennett, "The Baldwin Effect: A crane, not a skyhook" (Paper available at: ase.tufts.edu/cogstud/papers/baldwin-cranefin.htm).

See also Bruce H. Weber & David J. Depew, "Evolution and Learning" (Massachusetts Institute of Technology, 2003, which explores the Baldwin effect in detail).

12 Andrew Solomon, "The Autism Rights Movement" (*New York*, 25 May 2008, nymag.com/news/features/47225/).

See also Dr. Thomas Armstrong, *Neurodiversity: Discovering the extraordinary gifts of autism, ADHD, dyslexia, and other brain differences* (Da Capo Lifelong Books, May 2010).

See also Dr. Thomas Armstrong, "Other Forms of Neuro-diversity" (blog, 10 February 2010, thehumanodyssey.typepad. com/neurodiversity_the_book/other-forms-of-neurodiversity/).

13  Harvey Blume, "Neurodiversity" (*Atlantic*, 30 September 1998, www.theatlantic.com/magazine/archive/1998/09/neuro diversity/305909/).

14  "Response to the American Psychiatric Association: DSM-5 Development" (The British Psychological Society, apps.bps. org.uk/_publicationfiles/consultation-responses/DSM-5%20 2011%20-%20BPS%20response.pdf).

See also Lindsey Tanner, "Is Grief an Illness? DSM-5 editors face backlash over classification of 'normal life' as mental disorders" (*National Post*, 15 May 2013, life.nationalpost. com/2013/05/15/is-griefan-illness-dsm-5-editors-face-backlash-over-classification-ofnormal-life-as-mental-disorders/).

15  Harvey Blume, "Neurodiversity" (*Atlantic*, 30 September 1998, www.theatlantic.com/magazine/archive/1998/09/neuro diversity/305909/).

16  Laurent Mottron, "The Power of Autism" (*Nature*, 3 November 2011).

17  Simon Baron-Cohen, "The Extreme Male Brain Theory of Autism" (*Trends in Cognitive Sciences*, June 2002, cogsci.bme. hu/~ivady/bscs/read/bc.pdf).

18  Steve Silberman, "The Geek Syndrome" (*Wired*, issue 9, 12 December 2001, www.wired.com/wired/archive/9.12/aspergers. html).

19  Simon Baron-Cohen, "The Extreme Male Brain Theory of Autism" (*Trends in Cognitive Sciences*, June 2002, cogsci.bme. hu/~ivady/bscs/read/bc.pdf).

20  Damian Milton, "So What Exactly is Autism?" (Autism Education Trust, 2012, www.aettraininghubs.org.uk/wpcontent/ uploads/2012/08/1_So-what-exactly-is-autism.pdf).

21  Thomas Edison home page. www.thomasedison.com/index. html

22  Interview with Rollo Carpenter, June 2012.

23  David Shenk, *The Genius in All of Us* (Icon Books, 2011).

24 Gail Davies et al., "Genome-wide Association Studies Establish that Human Intelligence is Highly Heritable and Polygenic" (*Molecular Psychiatry*, October 2011).

25 Scott Barry Kaufman, "The Truth about the 'Termites'" (*Psychology Today*, 9 September 2009, www.psychologytoday.com/blog/beautiful-minds/200909/the-truth-about-the-termites).

    See also Mitchell Leslie, "The Vexing Legacy of Lewis Terman" (*Stanford*, August 2000, alumni.stanford.edu/get/page/magazine/article/?article_id=40678).

    See also Malcolm Gladwell, *Outliers: The Story of Success* (Penguin, 2008, pp. 74–75, 111–13).

26 Gregory Cochran & Henry Harpending, *The 10,000 Year Explosion* (Basic Books, 2010, p. 191).

27 David Brooks, "The Tel Aviv Cluster" (*New York Times*, 11 January 2011, www.nytimes.com/2010/01/12/opinion/12brooks.html?_r=0).

28 Paul Graham, "How to be Silicon Valley" (May 2006, www.paulgraham.com/siliconvalley.html).

29 New Economic Impact Study released. 1 April, 2014. High Tech Meetup. (http://nytm.org).

30 April Dembosky, "Over the Rainbow" (*Financial Times*, 29 June 2012, www.ft.com/intl/cms/s/2/091e540a-baf5-11e1-81e000144feabdc0.html#axzz28gZsk5O1).

31 Lawrence Lessig, "Sorkin vs. Zuckerberg" (*New Republic*, 1 October 2010, ww.tnr.com/article/books-and-arts/78081/sorkin-zuckerberg-the-social-network#).

32 James Gleick, "What Defines a Meme?" (Smithsonian.com, May 2011, www.smithsonianmag.com/arts-culture/What-Defines-aMeme.html).

33 C. J. Fuller & Haripriya Narasimhan, "Traditional Vocations and Modern Professions among Tamil Brahmans in Colonial and Post-colonial South India" (*Indian Economic Social History Review*, 2010, vol. 47, p. 473).

34 Darwin may have deliberately overstated this timeframe to help overcome potential incredulity about his radical theory. A longer timeframe made evolution seem less threatening.

35 Quoted in Michael Le Page, "Evolution in the Fast Lane: Unnatural selection" (*New Scientist*, 2 April 2011).

36 Quoted in Tony Fitzpatrick, "Does Environment Influence Genes? Researcher gives hard thoughts on soft inheritance" (Newsroom, Washington University in St. Louis, 3 August 2006, news.wustl.edu/news/Pages/7408.aspx).

37 Interview with Dr. Cath Suter, October 2012.
    See also J. E. Cropley, et al., "A Sustained Dietary Change Increases Epigenetic Variation in Isogenic Mice" (*PLOS Genetics*, 21 April 2011, www.plosgenetics.org/article/info%3Adoi%2F10.1371%2Fjournal.pgen.1001380).

38 John Cloud, "Why Your DNA Isn't Your Destiny" (*Time*, 6 January 2010, www.time.com/time/magazine/article/0,9171,1952313,00.html).

39 Dr. Tim Spector, "Identically Different" (blog, www.timspector.co.uk/?p=90).

40 Kara Rogers, "Epigenetics: A turning point in our understanding of heredity" (*Scientific American* blog, 16 January 2011, blogs.scientificamerican.com/guest-blog/2012/01/16/epigenetics-aturning-point-in-our-understanding-of-heredity/).

41 Interview with Rob Brooks, Professor of Evolutionary Biology, University of NSW, August 2012.

42 Benjamin Wallace, "Are you on it?" (*New York*, 28 October 2012, nymag.com/news/features/autismspectrum-2012-11/).

43 Quoted in Alice Park, "Autism Rises: More children than ever have autism, but is the increase real?" (*Time*, 29 March 2012, healthland.time.com/2012/03/29/autism-rises-more-u-children-than-ever-have-autism-is-the-increase-real/).

44 Dr. Thomas Insel, "The New Genetics of Autism: Why environment matters" (National Institute of Mental Health, www.nimh.nih.gov/about/director/2012/the-new-genetics-ofautism-why-environment-matters.shtml).

45 Juan Enriquez, "The Next Species of Human" (TED, April 2012, www.ted.com/talks/juan_enriquez_shares_mindboggling_new_science.html).

46 Steve Silberman, "The Geek Syndrome" (*Wired*, issue 9, 12 December 2001, www.wired.com/wired/archive/9.12/aspergers.html).

## CHAPTER THREE:
## THE RISE OF THE GEEK

1 The Rise of the geek quoted in Alexandra Petri, "Miss USA, Best Buy and the Geek Bubble" (*Washington Post*, 20 June 2011, www.washingtonpost.com/blogs/compost/post/miss-usa-best-buy-and-the-geek-bubble/2011/03/03/AGpDlJdH_blog.html).

2 Lauren Naefe, "Happy Geek Pride Day" (*Open Road*, 25 May 2011, www.openroadmedia.com/blog/tag/Glee.aspx).

3 Haya El Nasser, "Geek Chic: 'Brogrammer?' Now, that's hot" (*USA Today*, 12 April 2012, usatoday30.usatoday.com/tech/news/story/2012-04-10/techie-geeks-cool/54160750/1).

4 "Modis Geek Pride Day Survey Reveals Majority of Americans Believe Being Called a 'Geek' is a Compliment" (*Modis*, 23 May 2011, www.modis.com/it-insights/press-room/article/?art=20110523_1&type=pr).

5 James Ball, "Geekdom's a Broad Church – Q is Welcome at Our Door" (*Guardian*, 2 November 2012, www.guardian.co.uk/commentisfree/2012/nov/02/geekdom-broad-church-q-welcomejames-bond).

6 "Brits Wear Specs to Impress" (College of Optometrists, 5 January 2011, www.college-optometrists.org/en/knowledge-centre/news/index.cfm/Brits%20wear%20specs%20to%20impress).

7 "Wearing Glasses Can Improve Job Prospects" (*Telegraph*, 3 January 2011, www.telegraph.co.uk/finance/jobs/8237606/Wearing-glasses-can-improve-job-prospects.html).

8 Quoted in Haya El Nasser, "Geek Chic: 'Brogrammer?' Now, that's hot" (*USA Today*, 12 April 2012, usatoday30.usatoday.com/tech/news/story/2012-04-10/techie-geeks-cool/54160750/1).

See also Anand Giridhardada, "Silicon Valley Roused by Secession" (*New York Times*, October 28, 2013).

9 Benjamin Wallace, "Are you on it?" (*New York*, 28 October 2012, nymag.com/news/features/autismspectrum-2012-11/).

10 Steve Silberman, "What kind of Buddhist was Steve Jobs really?" (PLOS blogs, 28 October 2011, blogs.plos.org/neu rotribes/2011/10/28/what-kind-of-buddhist-was-steve-jobsreally/).

11 Richard von Sturmer, "Haiku and Zen" (New Zealand Poetry Society Te Hunga Tito Ruri o Aotearoa, www.poetrysociety. org.nz/node/113).

12 "Steve Jobs Speaks Out" (CNN Money, 7 March 2008, money.cnn.com/galleries/2008/fortune/0803/gallery.jobsqna. fortune/6.html).

13 Philip Elmer DeWitt, "Fortune names Jobs the greatest entrepreneur" (CNN Money, 25 March 2012, tech.fortune. cnn.com/2012/03/25/fortune-names-steve-jobs-the-greatest entrepreneur/)

14 Jay Cocks, "The Updated Book of Jobs" (*Time*, 3 January 1983).

15 Phillip Adams, "One Rotten Apple" (*The Australian*, 21 July 2012).

16 "'You've Got to Find What You Love,' Jobs says" (Stanford Report, 12 June 2005, news.stanford.edu/news/2005/june15/ jobs-061505.html).

17 Andrew G. Haldane, "Rethinking the Financial Network" (April 2009, see p. 10 of report: www.bankofengland.co.uk/ publications/Documents/speeches/2010/speech445.pdf).

18 Linda Davies, *Into the Fire* (Twenty-First Century Publishers, 2007).

19 Bank for International Settlements (BIS), www.bis.org/
See also New York Stock Exchange figures, www.nyse.com/

20 Federal Reserve Bank of New York, www.newyorkfed.org/

21 Quoted in Clive Cookson, Gillian Tett & Chris Cook, "Organic Mechanics" (*Financial Times*, 26 November 2009).

22 Albert Einstein, *Ideas and Opinions* (Bonanza Books, 1988).

23 Paul Kedrosky, "The First Disaster of the Internet Age" (*Newsweek*, 27 October 2008).

24 Presentation by Andrew G. Haldane, "Patience and Finance" (delivered to the Oxford China Business Forum, Beijing, on 9 September 2010, www.bankofengland.co.uk/publications/Documents/speeches/2010/speech445.pdf).

25 Andrew G. Haldane, "Rethinking the Financial Network" (April 2009, www.bis.org/review/r090505e.pdf).

26 David W. Orr, "Speed" (*Conservation Biology*, February 1998, pp. 4–7, www.jstor.org/stable/2387456).

27 Lawrence H. Summers, "Why American Families and Businesses Need Financial Reform" (*The White House Blog*, 15 October 2009, www.whitehouse.gov/blog/Why-American-Families-andBusinesses-Need-Financial-Reform).

28 Rana Foroohar, "Boom and Gloom" (*Newsweek*, 9 November 2009).

See also Reshmaan N. Hussam, David Porter & Vernon L Smith, "Thar She Blows: Can bubbles be rekindled with experienced subjects?" (*American Economic Review*, June 2008, people.exeter.ac.uk/dgbalken/BEEM10908/smith.pdf).

29 Melissa Rollock, "Before Google Ever Was" (Nationnews.com, 26 October 2008, archive.nationnews.com/archive_detail.php?archiveFile=2008/October/26/LocalNews/66971.xml&archive_pubname=Daily+Nation%0A%09%09%09%09).

30 Paul Gilster, *Digital Literacy* (Wiley, 1997, p. 148).

31 Stephanie Strom, "Billionaire Aids Charity that Aided Him" (*New York Times*, 24 October 2009, www.nytimes.com/2009/10/25/us/25donate.html?_r=).

32 Nicholas D. Kristof, "The Daily Me" (I, 18 March, 2009, www.nytimes.com/2009/03/19/opinion/19kristof.html?_r=0).

33 Bill Bishop, *The Big Sort: Why the Clustering of Like-minded America Is Tearing Us Apart* (Mariner Books, 2009, p 162).

34 Google code of conduct, investor.google.com/corporate/code-ofconduct.html

35 Leo Kelion, "GCHQ-backed Competition Names Cyber Security Champion" (BBC, 11 March 2012, www.bbc.com/news/technology-17333601).

36 Ibid.

37 "UK Cyber Crime Costs £27bn a Year–Government report" (BBC 17 February 2011, www.bbc.co.uk/news/ukpolitics-12492309).

38 Elisabeth Bumiller & Thom Shanker, "Panetta Warns of Dire Threat of Cyberattack on U.S." (*The New York Times*, 11 October 2012, www.nytimes.com/2012/10/12/world/panetta-warns-ofdire-threat-of-cyberattack.html?pagewanted=all).

39 Nick Hopkins, "US and China Engage in Cyber War Games" (*Guardian*, 16 April 2012, www.guardian.co.uk/technology/2012/apr/16/us-china-cyber-war-games).

40 Steinberg, Joseph, "Massive Internet Security Vulnerability – Here's What You Need to Do." (*Forbes*, 10 April 2014, http://www.forbes.com/sites/josephsteinberg/2014/04/10/massive-internet-security-vulnerability-you-are-at-risk-what-you-need-to-do/)

41 "Portrait of J. Random Hacker" (The Cyberpunk Project, 11 November 2003, project.cyberpunk.ru/idb/portrait_of_j_random_hacker.html).

42 Suelette Dreyfus, "Underground: Tales of hacking, madness and obsession on the electronic frontier" (*Mandarin Australia*, June 1997).

43 Gabriella Coleman, "Geeks Are the New Guardians of Our Civil Liberties" (*MIT Technology Review*, 4 February 2013, www.technologyreview.com/news/510641/geeks-are-the-newguardians-of-our-civil-liberties/).

44 Mike Sullivan, Anthony France, Neil Syson & Rhodri Phillips, "Essex Geek is Sony Hacker" (*Sun*, 22 June 2011, www.thesun.co.uk/sol/homepage/news/3651298/Essex-geek-Ryan-Cleary-isSony-hacker.html).

45 Roy Carroll, "Gary McKinnon Hacking Prosecution Called "Ridiculous" by US Defence Expert" (*Guardian*, 10

July 2012, www.guardian.co.uk/world/2012/jul/10/gary-mckinnonhacking-prosecution-us).

46 Nick Hopkins, "US and China Engage in Cyber War Games" (*Guardian*, 16 April 2012, www.guardian.co.uk/technology/2012/apr/16/us-china-cyber-war-games).

47 Alok Jha, "Underground River "Rio Hamza" Discovered 4km Beneath the Amazon" (*Guardian*, 26 August 2011, www.guardian.co.uk/environment/2011/aug/26/underground-riveramazon).

48 Donald Melanson, "Amazon Announces Q4 2011 Results: Sales jump to $17.43 billion, but profits drop 58 percent" (*Engadget*, 31 January 2012, www.engadget.com/2012/01/31/amazonannounces-q4-2011-results-sales-jump-to-17-43-billion/).

49 "AWS Case Study: Obama for America" (Amazon Web Services Case Study, aws.amazon.com/solutions/case-studies/obama/).

50 Richard L. Brandt, *One Click: Jeff Bezos and the Rise of Amazon.com* (Portfolio/Penguin, 2012).

51 "Jeffrey P. Bezos" (Academy of Achievement, 7 February 2013, www.achievement.org/autodoc/page/bez0bio-1).

52 Steven Levy, "Jeff Bezos Owns the Web in More Ways Than You Think" (*Wired*, 13 November 2011, www.wired.com/magazine/2011/11/ff_bezos/).

53 Jeff Bezos, "Baccalaureate remarks to Princeton University on 30 May 2010" (www.princeton.edu/main/news/archive/S27/52/51O99/index.xml).

54 Richard Russo, "Amazon's Jungle Logic" (*New York Times*, 12 December 2011, www.nytimes.com/2011/12/13/opinion/amazons-jungle-logic.html?pagewanted=all).

55 Elisabeth Bumiller, "A Day Job Waiting for a Kill Shot a World Away" (*New York Times*, 29 July 2012, www.nytimes.com/2012/07/30/us/drone-pilots-waiting-for-a-kill-shot-7000miles-away.html?pagewanted=all).

56 "F35: The Last Manned Fighter Jet" (*CNN Money*, 18 December 2013, https://www.youtube.com/watch?v=iAiSi1Eu1BA).

See also "The last manned fighter" (*Economist*, 14 July 2011, http://www.economist.com/node/18958487).

See also Peter L. Hartley, "The F-35—the UK's Last Manned Combat Aircraft Procurement?" (*Defense Update*, 18 March, 2012, http://defense-update.com/20120318_the-f-35-the-uks-last-manned-combat-aircraft-procurement.html#.U2ETYVVdWSo).

57  Patrick B. Pexton, "Are Drone Strikes the Only Game in Town?" (*National Journal*, 11 January 2010, security.nationaljournal.com/2010/01/are-drone-strikes-the-only-gam.php).

58  David Kilcullen & Andrew McDonald Exum, "Death From Above, Outrage Down Below" (*The New York Times*, 16 May 2009, www.nytimes.com/2009/05/17/opinion/17exum.html?pagewanted=all&_r=0).

59  Sebastian Anthony, "Just How Big Are Porn Sites?" (ExtremeTech, www.extremetech.com/computing/123929-justhow-big-are-porn-sites).

60  Benjamin Wallace, "The Geek Kings of Smut" (*New York*, 30 January 2011, nymag.com/news/features/70985/).

61  Chris Morris, "Meet the New King of Porn" (CNBC, 18 January 2012, www.cnbc.com/id/45989405/Meet_the_New_King_of_Porn).

62  Melinda Tankard Reist, "Self Love" (*The Drum*, 18 March 2010).

63  Gary Wilson & Marnia Robinson, "How Porn Can Ruin Your Sex Life, the Good Men Project" (8 February, 2011, goodmenproject.com/health/how-porn-can-ruin-your-sex-lifeand-your-marriage/).

64  Carla Marinucci, "Obama Dines with Tech Stars of Silicon Valley" (*SFGate*, 22 May 2013, www.sfgate.com/bayarea/article/Obama-dines-with-tech-stars-of-Silicon-Valley-2529458.php#ixzz2TvWxsbo5).

65  "Obama's Geeks v GOP Billions" (*Sydney Morning Herald*, 19 February 2012, www.smh.com.au/technology/technology-news/obamas-geeks-v-gop-billions-20120218-1tfu6.html).

66  Glenda Kwek, "Jim Messina: The man behind Obama's campaign" (Stuff.co.nz, 8 November 2012, www.stuff.co.nz/

world/americas/7926020/Jim-Messina-The-man-behind
Obamas-campaign).

67 Elliott Suthers, "Why the Geeks Are Ruining Politics" (*Forbes*,
15 March 2012, www.forbes.com/sites/elliottsuthers/2012/03/
15/why-the-geeks-are-ruining-politics/).

68 Alexander Burns, "Romney Makes Birth Certificate Joke"
(Politico.com, 24 August 2012, www.politico.com/blogs/
burnshaberman/2012/08/romney-makes-birth-certificate-
joke-133091.html).

69 Andrew Hough, "Nate Silver: Politics 'geek' hailed for Barack
Obama wins US election forecast" ((UK) *Telegraph*, 7 November
2012, www.telegraph.co.uk/news/worldnews/uselection/9662363/
Nate-Silver-politics-geek-hailed-for-BarackObama-wins-US-
election-forecast.html).

70 Ibid.

71 Ibid.

72 Mark Henderson, *The Geek Manifesto: Why Science Matters*
(Bantam Press, 2012).

73 Tom Peck, "Ed Miliband: The geek in politics" (*Independent*,
6 October 2012, www.independent.co.uk/news/uk/politics/
ed-miliband-the-geek-in-politics-8200240.html).

74 "Isaac Newton (1643-1727)" (BBC, www.bbc.co.uk/history/
historic_figures/newton_isaac.shtml).

75 Jimmy Carter *Playboy* interview (PBS, www.pbs.org/news-
hour/character/glossaries/carter.html).

76 Suzanne Moore, "The end of spin? Don't be daft. We've for-
gotten how to do without it" (*Guardian*, 22 July 2011, www.
guardian.co.uk/media/2011/jul/22/end-spin-dont-be-daft).

77 Lucy Ash, "South Korea's e-sports stars" (3 January, 2008.
news.bbc.co.uk/2/hi/programmes/crossing_continents/
7167890.stm).

78 "The new reality: Call of Duty beats Avatar sales record"
(*Independent*, 13 December 2011, www.independent.co.uk/
lifestyle/gadgets-and-tech/news/the-new-reality-call-
of-duty-beatsavatar-sales-record-6276139.html).

79 Alex Rayner, "Are Video Games Just Propaganda and Train-ing Tools for the Military?" (*Guardian*, 18 March 2012, www.guardian.co.uk/technology/2012/mar/18/video-gamespropaganda-tools-military).

80 Interview by Mark Roeder with hosts of ABC *Good Game Spawn Point*, August 2012, at ABC Studios.

81 Jefferson Graham, "Wikipedia: Jimmy Wales takes wiki work seriously" (*USA Today*, 11 May 2011, usatoday30.usatoday.com/tech/news/2011-05-10-wikipedia-jimmy-wales_n.htm).

82 Interview by Mark Roeder with hosts of ABC *Good Game Spawn Point*, August 2012, at ABC Studios.

83 Peter Gray, "The Many Benefits, for Kids, of Playing Video Games" (*Psychology Today*, 7 January 2012, www.psy-chologytoday.com/blog/freedom-learn/201201/the-many benefits-kids-playing-video-games).

84 Tom Bissell, *Extra Lives: Why Video Games Matter* (Vintage, 2011).

85 Interview with Max Fulcher, June 2012.

86 Kathryn Zickuhr & Mary Madden, "Older Adults and Internet Use" (Pew Internet Research, 6 June 2012, pewinternet.org/Reports/2012/Older-adults-and-internet-use/Main-Report/Gadget-ownership.aspx).

87 Ben Sisaro, "Susan Boyle, Top Seller, Shakes Up CD Trends" (*New York Times*, 2 December 2009, www.nytimes.com/2009/12/03/arts/music/03sales.html?_r=0).

88 Jose Antonio Vargas, "Scottish Singer Susan Boyle's Web Pop-ularity Is at Numbers Never Seen Before" (*Washington Post*, 20 April 2009, www.washingtonpost.com/wp-dyn/content/article/2009/04/19/AR2009041900508.html).

89 Abby Stokes, *Is This Thing On? A Computer Handbook for Late Bloomers, Technophobes and the Kicking and Screaming* (Workman, 2008. p. 74).

90 "'App' Voted 2010 Word of the Year by the American Dia-lectic Society" (American Dialectic Society, 8 January 2011,

www.americandialect.org/app-voted-2010-word-of-the-year-by-theamerican-dialect-society-updated).

91 Somini Sengupta, Nicole Perroth & Jenna Wortham, "Behind Instagram's Success: Networking the old way" (*New York Times*, 10 April 2012, www.nytimes.com/2012/04/14/technology/instagram-founders-were-helped-by-bay-areaconnections.html?pagewanted=all).

92 Chris Foresman, "iOS App Success is a 'Lottery': 60% (or more) of developers don't break even" (*ars technica*, 5 May 2012, arstechnica.com /apple/2012/05/ios-app-success-is-a-lottery-and60-of-developers-dont-break-even/).

93 Tom Chatfield, "When Smart is not so Smart" (*Future*, BBC, 1 February 2013, www.bbc.com/future/story/20130201-whensmart-is-not-so-smart).

## CHAPTER FOUR:
## THE COGNITIVE REVOLUTION

1 J. Savulescu, N. Bostrom, B. Sahakian et al. "Cognitive Enhancement" (Oxford Centre for Neuroethics, www.neuroethics.ox.ac.uk/research/area_1).

2 Norman Doidge, *The Brain That Changes Itself: Stories of Personal Triumph from the Frontiers of Brain Science* (Penguin, 2007).

3 Sharon Begley, *Train Your Mind, Change Your Brain: How a New Science Reveals Our Extraordinary Potential to Transform Ourselves* (Ballantine Books, 2007).

4 William J. Cromie, "Meditation Found to Increase Brain Size" (*Harvard Gazette*, 2 February 2006, news.harvard.edu/gazette/story/2006/02/meditation-found-to-increase-brain-size/).

5 Jon Kabat-Zinn, *Wherever You Go, There You Are: Mindfulness Meditation for Everyday* (Piatkus, 2004).

6 Maia Szalavitz, "Popping Smart Pills: The Case for Cognitive Enhancement" (*Time*, 6 January 2009, www.time.com/time/health/article/0,8599,1869435,00.html).

7   Philippa Roxby, "Do 'Smart Drugs' Really Make Us Brain-
    ier" (*BBC News*, 3 April 2011, www.bbc.co.uk/news/health-
    12922451).

8   Alan Schwarz, "Risky Rise of the Good-Grade Pill" (*New York
    Times*, 9 June 2012, www.nytimes.com/2012/06/10/education/
    seeking-academic-edge-teenagers-abuse-stimulants.
    html?pagewanted=all&_r=0).

9   Steve Boggan & Tim Stewart, "Illegal 'Smart Drugs' Bought
    Online by Teenagers Before Exams Could Have Catastrophic
    Effect on Their Health" (*Daily Mail*, 10 March 2010, www.
    dailymail.co.uk/health/article-1256481/ Illegal-smart-drugs-
    bought-online-teenagers-exams-catastrophic-effect-health.
    html).

10  Maia Szalavitz, "Popping Smart Pills: The Case for Cognitive
    Enhancement" (*Time*, 6 January 2009, www.time.com/time/
    health /article/0,8599,1869435,00.html).

11  Tom Newham, "Smart Drugs: Would you try them?" (*Guard-
    ian*, 24 October 2012, www.guardian.co.uk/education/mortar
    board/2012/oct/24/smart-drugs-would-you-try-them).

12  Alan Schwarz, "Risky Rise of the Good-Grade Pill" (*New
    York Times*, 9 June 2012, www.nytimes.com/2012/06/10/edu-
    cation/seeking-academic-edge-teenagers-abuse-stimulants.
    html?pagewanted=all&_r=0).

13  Donald W. Light, *The Risks of Prescription Drugs* (Columbia
    University Press, 2010).

14  Interview with Rob Brooks by the author, Sydney, October
    2012.

15  Henry Greely et al., "Towards Responsible Use of Cognitive
    Enhancing Drugs by the Healthy" (*Nature*, 7 December 2008,
    npp.wisc.edu/newsarchive/PDF/TowardsResponsibleU-
    seOFCognitiveEnhancingDrugsByTheHealthy.pdf).

16  Maia Szalavitz, "Popping Smart Pills: The case for cognitive
    enhancement" (*Time*, 6 January 2009, www.time.com/time/
    health/article/0,8599,1869435,00.html).

17  Interview with Rob Brooks, Sydney, October 2012.

18 "Enlightenment Man" (*Economist*, 4 December 2008, www. economist.com/node/12673407).

19 "Scripps Research Institute Study Suggests Expanding the Genetic Alphabet May Be Easier than Previously Thought" (news release from the Scripps Research Institute, 3 June 2012, www.scripps.edu/news/press/2012/20120603romesberg.html).

20 Daniel H. Wilson, *Amped: A Novel* (Doubleday, June 2012).

21 Daniel H. Wilson, "Bionic Brains and Beyond" (*Wall Street Journal*, 1 June 2012, online.wsj.com/article/SB100014240527 02303640104577436601227923924.html).

22 "Successes of Deep Brain Stimulation for Patients with Parkinson's Disease" (Institute of Neurology, 13 April 2011, www.ucl.ac.uk/ion/articles/news/functional-neurosurgery-deepbrain-stimulation-parkinsons-foltynie).

23 Sarah Fox, "Can Neural Implants Hotwire Damaged Brain Circuits?" (The Brain Bank, 4 October 2012, thebrainbank. scienceblog.com/2012/10/04/could-neural-implants-be-usedto-hotwire-damaged-brain-circuits/).

  See also Douglas Heaven, "Neural Implants Could Spark Better Decisions" (*New Scientist*, 19 September 2012).

24 Daniel H. Wilson, "Bionic Brains and Beyond" (*Wall Street Journal*, 1 June 2012, online.wsj.com/article/SB100014240527 02303640104577436601227923924.html).

25 Nick Bilton, "Disruptions: Next Step for Technology Is Becoming the Background" (*New York Times*, 1 July 2012, bits.blogs.297    nytimes.com/2012/07/01/google's-project-glass-lets-technology slip-into-the-background/).

26 Ibid.

27 Charles Arthur, "Google Glass: Is it a threat to our privacy?" (*Guardian*, 6 March 2013, www.guardian.co.uk/technology/ 2013/mar/06/google-glass-threat-to-our-privacy).

28 Daniel C. Dennett, "A Perfect and Beautiful Machine: What Darwin's Theory of Evolution Reveals about Artificial Intelligence" (*Atlantic*, 22 June 2012, www.theatlantic. com/technology/archive/2012/06/a-perfect-and-beautiful-

machine-what-darwinstheory-of-evolution-reveals-about-artificial-intelligence/258829/).

29 John Markoff, "How Many Computers to Identify a Cat? 16,000" (*New York Times*, 25 June 2012, www.nytimes. com /2012/06/26/technology/in-a-big-network-of-computersevi-dence-of-machine-learning.html?pagewanted=all&_r=0).

See also Quoc V. Le et al., "Building High-level Features Using Large-Scale Unsupervised Learning" (International Conference on Machine Learning, 2012, static.googleusercontent. com/external_content/untrusted_dlcp/research.google.com/en/us/pubs/archive/38115.pdf).

30 Julian Sanchez, *Twitter*, 26 June 2012, https://twitter.com/normative/status/217679576984334336

31 Marcus du Sautoy, "AI Robot: how machine intelligence is evolving" (*Guardian*, 1 April 2012, www.guardian.co.uk/tech nology/2012/apr/01/ai-artificial-intelligence-robots-sautoy).

CHAPTER FIVE:
TRANSCENDENCE

1 Rhett Butler, "Epiphytes" (Mongabay.com, 30 July 2012, rain-forests.mongabay.com/0405.htm).

2 Follow-up communications with Stuart Pearson by the author in July 2012 and May 2013.

3 Ray Kurzweil, *The Singularity is Near: When Humans Transcend Biology* (Penguin, 2006).

4 Pierre Teilhard de Chardin, *The Future of Man* (Image, 2004).

5 Julian Huxley, *New Bottles for New Wine* (Chatto & Windus, 1957, pp. 13–17).

6 FM-2030, *Are You a Transhuman? Monitoring and Stimulating Your Personal Rate of Growth in a rapidly changing world* (Warner Books, January 1989).

See also Douglas Martin, "Futurist Known as FM-2030 Is Dead at 69" (*New York Times*, 11 July 2000, www.nytimes.com/2000/07/11/us/futurist-knownas-fm-2030-is-dead-at-69.html).

7   Ronald Bailey, *Liberation Biology: The scientific and moral case for the biotech revolution* (Prometheus Books, 2005).

8   John Harris, "Who's Afraid of a Synthetic Human?" (*Times*, 17 May 2008).

    See also www.almendron.com/tribuna/who'safraid-of-a-synthetic-human/

## CHAPTER SIX:
## BACKLASH

1   "Superhuman" exhibition, London, July-October 2012, www.wellcomecollection.org/whats-on/exhibitions/superhuman.aspx

2   Francis Fukuyama, "Transhumanism" (*Foreign Policy*, 1 September 2004, www.foreignpolicy.com/articles/2004/09/01/transhumanism).

3   Gerald F. Joyce, "Toward an Alternative Biology" (*Science*, 20 April 2012, pp. 307–8, www.sciencemag.org/content/336/6079/307).

4   Tony Bartlett, "Humans Become 'Pets' in Rise of the Machines: Apple co-founder" (*The Age*, 6 June 2011, www.theage.com.au/technology/technology-news/humans-becomepets-in-rise-of-the-machines-apple-cofounder-20110603-1fkqo.html#ixzz2TviV THpp).

5   Asher Moses, "Rise of the Machines" (*Sydney Morning Herald*, 18 July 2012, www.smh.com.au/technology/sci-tech/rise-of-themachines-20120718-229ev.html).

6   Peter A. Levine, *Waking the Tiger: Healing Trauma: The Innate Capacity to Transform Overwhelming Experiences* (North Atlantic Books, 1997).

    See also Peter A. Levine, "Healing Trauma: A pioneering program for restoring the wisdom of your body" (*Sounds True*, October 2008).

7   Jon Kabat-Zinn, *Wherever You Go, There You Are* (Hyperion, 2005).

8 Kevin Binfield, *Writings of the Luddites* (Johns Hopkins University Press, 2004).

See also Richard Conniff, "What the Luddites Really Fought Against" (Smithsonian.com, March 2012, www.smithsonianmag. com/history-archaeology/Whatthe-Luddites-Really-Fought-Against.html).

9 "Wingspread Conference on the Precautionary Principle" (Science and Environmental Health Network, 26 January 1998, www.sehn.org/wing.html).

10 William Safire cited in "Explore a Topic: Neuroethics" (Bioethics Research Library Georgetown University, bioethics. georgetown.299edu/resources/topics/neuroethics/index.html).

11 International Neuroethics Society, www.neuroethicssociety.org

12 Oxford Centre for Neuroethics, (Wellcome Trust), www.neuroethics.ox.ac.uk

13 Rietberg Museum, Zurich, Switzerland, www.rietberg.ch/de-ch/foyer.aspx

14 George Bernard Shaw, *The Perfect Wagnerite* (IndyPublish, 2008).

15 Alex Ross, "Secret Passage: Decoding Ten Bars in Wagner's 'Ring'" (*The New Yorker*, 25 April 2011, www.newyorker. com/arts/critics/atlarge/2011/04/25/110425crat_atlarge_ross).

16 Michael Shermer, *The Believing Brain: From Ghosts and Gods to Politics and Conspiracies—How We Construct Beliefs and Reinforce Them as Truths* (St Martin's Griffin, 2012; see also www.michaelshermer.com/the-believing-brain/).

17 Jay Yarow, "Sergey Brin: We Want Google To Be the Third Half of Your Brain" (*Business Insider Australia*, au.businessinsider. com/sergey-brin-we-want-google-to-be-thethird-half-of-your-brain-2010-9).

18 Carl Sagan, *Pale Blue Dot: A Vision of the Human Future in Space* (Ballantine Books, 1997).

19 Edward O. Wilson, *The Social Conquest of Earth* (Liveright Publishing Company, 2012).

# Bibliography

Armstrong, Thomas. *Neurodiversity: Discovering the Extraordinary Gifts of Autism, ADHD, Dyslexia, and Other Brain Differences.* Cambridge, MA: Da Capo Lifelong Books, 2010.

Begley, Sharon. *Train Your Mind, Change Your Brain: How a New Science Reveals Our Extraordinary Potential to Transform Ourselves.* New York: Ballantine Books, 2007

Binfield, Kevin. *Writings of the Luddites.* Baltimore: The Johns Hopkins University Press, 2004.

Brooks, David. *The Social Animal: The Hidden Sources of Love, Character, and Achievement.* New York: Random House, January 2012.

Brooks, Rob. *Genes, Sex and Rock: How Evolution Has Shaped the Modern World.* Sydney: New South Books, 2011.

Darwin, Charles. *On the Origin of Species.* Cambridge: Harvard University Press (facsimile edition), 2001.

Dawkins, Richard. *The Blind Watchmaker.* New York: W. W. Norton & Company, 1997.

———. *The Selfish Gene.* Oxford: Oxford University Press, 1990.

Dennett, Daniel C. *Darwin's Dangerous Idea.* New York: Simon & Schuster, 1996.

Doidge, Norman. *The Brain That Changes Itself: Stories of Personal Triumph from the Frontiers of Brain Science.* New York: Penguin, 2007.

Drefus, Suelette. *Underground: Tales of Hacking, Madness, and Obsession on the Electronic Frontier.* Sydney: Mandarin Australia, 1997.

Gaddam, Sai & Ogas, Ogi. *A Billion Wicked Thoughts.* New York: Plume Books, 2012.

Gladwell, Malcolm. *Outliers: The Story of Success.* New York: Penguin, 2008.

Henderson, Mark. *The Geek Manifesto: Why Science Matters.* London: Bantam Press, 2012.

Kabat-Zinn, Jon. *Wherever You Go, There You Are: Mindfulness Meditation for Everyday.* London: Piatkus Books, 2004.

Kurzweil, Ray. *The Age of Spiritual Machines: When Computers Exceed Human Intelligence.* New York: Penguin, 2000.

———. *The Singularity Is Near: When Humans Transcend Biology.* New York: Penguin Books, 2006.

Levine, Peter A. *Waking the Tiger: Healing Trauma: The Innate Capacity to Transform Overwhelming Experiences.* Berkeley: North Atlantic Books, July 1997.

Pinker, Steven. *The Language Instinct: How the Mind Creates Language.* New York: Harper Perennial Modern Classics, 2007.

Sagan, Carl. *Pale Blue Dot: A Vision of the Human Future in Space.* New York: Ballantine Books, 1997.

Tett, Gillian. *Fool's Gold.* New York: Little, Brown, 2009.

Wilson, Daniel H. *Amped: A Novel.* New York: Doubleday, 2012.

Wilson, Edward O. *The Social Conquest of Earth.* New York: Liveright Publishing Company, 2012.

# Index